Progress in Mathematics
Vol. 20

Edited by
J. Coates and
S. Helgason

Birkhäuser
Boston · Basel · Stuttgart

Glenn Stevens

Arithmetic on Modular Curves

1982

Birkhäuser
Boston • Basel • Stuttgart

Author:

Glenn Stevens
Department of Mathematics
Rutgers University
New Brunswick, New Jersey 08903

Library of Congress Cataloging in Publication Data

Stevens, Glenn, 1953-
Arithmetic on modular curves.
(Progress in mathematics ; v. 20)
includes bibliographical references.
1. Forms, Modular. 2. Curves, Modular.
3. L-functions. 4. Congruences and residues.
I. Title. II. Series: Progress in mathematics
(Cambridge, Mass.) ; 20.
QA243.S77 512'.72 82-4306
AACR2

CIP-Kurztitelaufnahme der Deutschen Bibliothek

Stevens, Glenn:
Arithmetic on modular curves / Glenn Stevens.
-Boston ; Basel ; Stuttgart : Birkhäuser, 1982.
(progress in mathematics ; Vol.20)

NE: GT

ISBN 978-0-8176-3088-1 ISBN 978-1-4684-9165-4 (eBook)
DOI 10.1007/978-1-4684-9165-4

Dedicated to Mrs. Helen Hammitt

Table of Contents

Introduction

One of the most intriguing problems of modern number theory is to relate the arithmetic of abelian varieties to the special values of associated L-functions. A very precise conjecture has been formulated for elliptic curves by Birch and Swinnerton-Dyer and generalized to abelian varieties by Tate. The numerical evidence is quite encouraging. A weakened form of the conjectures has been verified for CM elliptic curves by Coates and Wiles, and recently strengthened by K. Rubin. But a general proof of the conjectures seems still to be a long way off.

A few years ago, B. Mazur [26] proved a weak analog of these conjectures. Let N be prime, and f be a weight two newform for $\Gamma_0(N)$. For a primitive Dirichlet character χ of conductor prime to N, let $\Lambda_f(\chi)$ denote the algebraic part of $L(f, \chi, 1)$ (see below). Mazur showed in [26] that the residue class of $\Lambda_f(\chi)$ modulo the "Eisenstein" ideal gives information about the arithmetic of $X_0(N)$. There are two aspects to his work: congruence formulae for the values $\Lambda_f(\chi)$, and a descent argument.

Mazur's congruence formulae were extended to $\Gamma_1(N)$, N prime, by S. Kamienny and the author [17], and in a paper which will appear shortly, Kamienny has generalized the descent argument to this case.

The purpose of this monograph is to describe generalizations of the congruence formulae to other congruence groups. The present work differs in two ways from the earlier approach. First of all, the modular units, which Mazur uses to produce his parabolic cohomology classes, are replaced by their logarithmic derivatives which are weight two Eisenstein series. This makes the congruence formulae more natural, and reduces lengthy calculations with Dedekind sums to well known results concerning special values of Dirichlet L-functions. Since Eisenstein series appear in many settings, we may also anticipate that these methods will generalize, e.g. to the higher weight case or to Hilbert modular forms. The second difference is that the decomposition of the cuspidal group in the factors of the modular Jacobian is used to determine which congruences are satisfied by special values of L-functions of which cusp forms.

Let Γ be a congruence group of type (N_1, N_2) and of level $N = \mathrm{lcm}(N_1, N_2)$ (see §1.1). Let X be the associated modular curve, f be a normalized weight two newform for Γ, and $A_f \subseteq \mathrm{Pic}^0(X)$ be the abelian subvariety associated to f ([42], Theorem 7.14). Let $\mathcal{O}(f) \subseteq \mathbb{C}$ be the ring generated by the eigenvalues of the Hecke operators acting on f. Let $\mathbb{T} \subseteq \mathrm{End}(\mathrm{Pic}^0(X)/\mathbb{Q})$ be the ring of endomorphisms generated by the Hecke operators. Then $\mathcal{O}(f)$ is identified with the ring of endomorphisms of A_f obtained by restriction of $\mathbb{T}$ to A_f. Let $\mathfrak{P} \subseteq \mathcal{O}(f)$ be a prime ideal and $k = \mathcal{O}(f)/\mathfrak{P}$. Let $A_f[\mathfrak{P}]$ denote the $\mathfrak{P}$-torsion subgroup of A_f. Then $A_f[\mathfrak{P}]$ is a k-vector space. Assume $\mathfrak{P}$ satisfies

$$(*) \qquad \dim_k A_f[\mathfrak{P}] = 2 \quad .$$

If A_f is an elliptic curve this condition is trivially satisfied. In general (*) holds for all but finitely many primes $\mathfrak{P}$ in $\mathcal{O}(f)$.

Let $\mathbb{T}$ act on $\mathbb{C}$ through the natural homomorphism $\mathbb{T} \to \mathcal{O}(f) \subseteq \mathbb{C}$. Then integration with respect to the differential form $\omega(f)$ on X whose pullback to $\mathcal{H}$ is $f(z)\,dz$ gives rise to a $\mathbb{T}$-homomorphism

$$\varphi_f : H_1(X;\mathbb{Z}) \longrightarrow \mathbb{C}$$
$$\gamma \longmapsto \int_\gamma \omega(f) \quad .$$

There are periods $\Omega_f^+, \Omega_f^- \in \mathbb{C}^*$ and $\mathbb{T}$-homomorphisms $\psi_f^+, \psi_f^- : H_1(X;\mathbb{Z}) \to \mathcal{O}(f)$ such that the images are not contained in $\mathfrak{P}$ and

$$\varphi_f(\gamma) = \psi_f^+(\gamma)\,\Omega_f^+ + \psi_f^-(\gamma)\,\Omega_f^- \quad .$$

Of course $\Omega_f^\pm$ are determined only up to multiplication by a $\mathfrak{P}$-unit in $\mathcal{O}(f)$.

THEOREM (Shimura [43]): Let χ be a nontrivial primitive Dirichlet character of conductor prime to N. Then

$$\Lambda_f(\chi) \overset{\text{dfn}}{=} \frac{\tau(\bar{\chi})\,L(f,\chi,1)}{2\pi i\,\Omega_f^{\operatorname{sgn}(\chi)}} \in \mathcal{O}(f)[\chi]$$

where $\tau(\bar{\chi}) = \sum_{a=1}^{m_\chi - 1} \bar{\chi}(a)\, e^{2\pi i a/m_\chi}$. □

We will refer to $\Lambda_f(\chi)$ as the algebraic part of $L(f, \chi, 1)$.

It is natural to reduce the values $\Lambda_f(\chi)$ modulo a prime $\mathfrak{P}$ of $\overline{\mathbb{Q}}$ extending $\mathfrak{p}$. By the Eichler-Shimura relation ([42], Theorem 7.9) we know that the Fourier coefficients $a_n(f)$, $(n, N) = 1$, are determined modulo $\mathfrak{p}$ by the Galois representation on $A_f[\mathfrak{p}]$. This motivates us to ask the following basic question.

<u>Question 1</u>: Does the group $A_f[\mathfrak{p}]$ determine the congruence class of $\Lambda_f(\chi)$ (modulo $\mathfrak{P}$) (up to multiplication by a $\mathfrak{P}$-unit independent of χ)? □

We might even ask the following.

<u>Question 2</u>: Can an explicit formula be given for $\Lambda_f(\chi)$ modulo $\mathfrak{P}$ in terms of the group $A_f[\mathfrak{p}]$? □

In the following pages we will answer Question 2 affirmatively in a special case, namely when $A_f[\mathfrak{p}]$ contains a nontrivial subgroup of the cuspidal divisor class group.

Let E be a weight two Eisenstein series for Γ which is an eigenfunction for the Hecke operators. Then there are Dirichlet characters $\varepsilon_1, \varepsilon_2$ of conductors N_1, N_2 respectively such that for all primes $\ell \nmid N$,

$$(E|T_\ell) = (\varepsilon_1(\ell) + \ell\,\varepsilon_2(\ell)) \cdot E \quad .$$

In §1.8 we will associate to E a subgroup, C_E, of the cuspidal group. In Chapter 3 we show that C_E is a cyclic $\mathbb{T}[\mathrm{Gal}(\overline{\mathbb{Q}}/\mathbb{Q})]$-module on which

$\mathrm{Gal}(\overline{\mathbb{Q}}/\mathbb{Q})$ acts via the character ϵ_1.

Suppose that in addition to (*) we also have

$$(**) \qquad C_E \cap A_f[\mathfrak{P}] \neq 0 \quad .$$

Then there is a Galois conjugate, E_σ, of E for which the eigenvalues of Hecke satisfy $a_n(E_\sigma) \equiv a_n(f) \pmod{\mathfrak{P}}$ for all $n > 0$ (Proposition 4.1.4). Without loss of generality we may assume $E_\sigma = E$.

Since $C_E \cap A_f[\mathfrak{P}]$ is a one dimensional k-subspace of $A_f[\mathfrak{P}]$ on which $\mathrm{Gal}(\overline{\mathbb{Q}}/\mathbb{Q})$ acts via ϵ_1, the Galois representation $\rho : \mathrm{Gal}(\overline{\mathbb{Q}}/\mathbb{Q}) \to GL_2(k) \cong \mathrm{Aut}(A_f[\mathfrak{P}])$ is given by

$$\rho(\mathrm{Frob}_\ell) = \begin{pmatrix} \epsilon_1(\ell) & * \\ 0 & \ell\,\epsilon_2(\ell) \end{pmatrix}$$

for ℓ prime to $N\mathfrak{P}$.

We will prove (Theorem 4.2.3) the following:

THEOREM: Assume (*) and (**). Then there is a complex number $\Omega_E \in \mathbb{C}^*$ such that for all nontrivial primitive Dirichlet characters χ with conductor prime to $N \cdot \mathfrak{P}$ and satisfying $\mathrm{sgn}(\chi) = -\mathrm{sgn}(\epsilon_1)$:

(1) $\Lambda_E(\chi) \overset{\text{dfn}}{=} \dfrac{\tau(\overline{\chi})\, L(E,\chi,1)}{2\pi i\, \Omega_E}$ is $\mathfrak{P}$-integral

and (2) $\Lambda_f(\chi) \equiv \Lambda_E(\chi)$ (modulo $\mathfrak{P}$). □

In §4.3 we consider the example $X = X_0(49)$. This is Gross's "$\mathbb{Q}$-curve" $A(7)$ [9]. It is an elliptic curve defined over $\mathbb{Q}$ and has CM by the integers of $K = \mathbb{Q}(\sqrt{-7})$. Let f be the unique normalized newform for $\Gamma_0(49)$. Let $\mathcal{O} = \mathbb{Z}[\rho]$, $\rho = e^{\pi i/3}$, and $\pi = 3 - \rho \in \mathcal{O}$ one of the primes dividing 7. Let $\varepsilon : \mathbb{Z} \to \mathbb{C}$ be the Dirichlet character of conductor 7 satisfying $\varepsilon(3) = \rho$. Then for each nontrivial even primitive Dirichlet character χ of conductor m_χ prime to 7 we have

$$\Lambda_f(\chi) \equiv \chi(7) \cdot \bar{\varepsilon}(m_\chi) \cdot \mathbf{B}_1(\varepsilon\chi) \cdot \mathbf{B}_1(\varepsilon\bar{\chi}) \pmod{\mathfrak{P}} \quad .$$

where $\mathfrak{P}$ is any prime of $\overline{\mathbb{Q}}$ extending (π).

This congruence was first observed experimentally by B. Gross. Recently, Gross provided a descent argument which shows that the congruence is compatible with the conjecture of Birch and Swinnerton-Dyer.

Fix an integer $m > 0$ prime to 7. By a result of E. Friedman [7] there are only finitely many primitive even Dirichlet characters χ of conductor dividing m^∞ for which the right-hand side of the above congruence is zero modulo $\mathfrak{P}$. This fact was used by K. Rubin [37] and A. Wiles to show that the group

$$X_0(49)\left(K \cdot \mathbb{Q}\left(\mu_{m^\infty}\right)^+\right)$$

is finitely generated. The finite generation also follows from Gross's descent.

We now give a rough sketch of the structure of the book. More detailed descriptions can be found at the beginning of each chapter.

Chapter 1 is devoted to describing the basic concepts which are used later. Of particular significance to us are the universal special value (§1.6), and the relation between the space of Eisenstein series and the cuspidal group (§1.8).

Chapter 2 studies periods of weight two modular forms $f \in \mathfrak{m}_2$ of all levels. We construct a cocycle

$$\pi \in Z^1(GL_2^+(\mathbb{Q}); \mathrm{Hom}_{\mathbb{C}}(\mathfrak{m}_2, \mathbb{C})$$

and describe it in terms of special values of L-functions. Restricting this cocycle to the space $\mathcal{E}_2$ of Eisenstein series and passing to the "real part" we obtain a "rational" cocycle

$$\xi \in Z^1(GL_2^+(\mathbb{Q}); \mathrm{Hom}_{\mathbb{C}}(\mathcal{E}_2, \mathbb{C}))$$

for which we give an explicit formula.

In Chapter 3 we use the cocycle ξ to associate a congruence for the universal special value to an arbitrary Eisenstein series. This is illustrated in §3.4 for a system of Eisenstein series for Γ which are eigenfunctions for the Hecke operators T_ℓ, $\ell \nmid N_1N_2$. In §3.5 we show how a theorem of Friedman and Washington can be used to prove nonvanishing results for special values of L-functions.

In Chapter 4 we define the Eisenstein primes. We translate the results of Chapter 3 into congruence formulae for the values $\Lambda_f(\chi)$ modulo an Eisenstein prime $\mathfrak{P} \subseteq \mathfrak{O}(f)$ satisfying (*) and (**). In §4.3 we give two examples: $X_0(7,7)$ $(\cong X_0(49))$, and $X_1(13)$.

In Chapter 5 we use the method of Mazur and Swinnerton-Dyer [30] to define two p-adic L-functions. The first of these is a p-adic version, $L_p(E, \chi, s)$, of the complex L-function attached to an Eisenstein series E. Typically $L_p(E, \chi, s)$ is a product of two Kubota-Leopoldt p-adic L-functions. The second of these is the P-adic L-function $L_P(X(\Gamma), \chi, s)$ of the modular curve $X(\Gamma)$ attached to an Eisenstein prime P for E. The main result (§5.6) is that these two p-adic L-functions are related by a congruence.

We have compiled in Chapter 6 a collection of tables of the algebraic parts of special values of L-functions attached to cusp forms f twisted by quadratic characters. The tables are for the following modular curves:

1. $X_0(N)$, N prime ≤ 43;
2. $X_0(N)$ of genus one;
3. $X_1(13)$, odd characters only.

Acknowledgements

It is with great pleasure that I give thanks to the many people whose friendship and support have been tremendously important to the writing of this manuscript.

I feel a unique sense of gratitude to Mrs. Helen Hammitt who, years ago, introduced me to the beauty of mathematics. Her strength and love are an inspiration to all who know her. I cannot give enough thanks to my parents, June and Earl Stevens. Their love and friendship have guided and protected me through the years.

Special thanks are due to Shelly who saved me from the modular symbols.

I am particularly indebted to Barry Mazur who got me started on the problems discussed in this volume. Joint work with Sheldon Kamienny on $X_1(p)$ pointed the way to the general situation. Stimulating conversations with friends and colleagues Bob Friedman, Bruce Jordan, Ron Livne, and Andrew Wiles have been lots of fun and very helpful.

Finally, I would like to express my appreciation to Barbara Moody who did a beautiful job of typing the manuscript, and also to Ruby Aguirre who patiently typed the tables.

Chapter 1. Background

We begin with a tour of the basic concepts which we will study in more detail in the following chapters. For our purposes the most important of these are the universal special value and the cuspidal group.

The universal special value was introduced by B. Mazur [26] as a means of understanding the algebraic nature of the special value $L(f, \chi, 1)$ attached to a weight two cusp form f and a Dirichlet character χ. It is defined to be a 1-cycle, $\Lambda(\chi)$, on a modular curve, X, on which f represents a holomorphic 1-form. By integrating f along this 1-cycle we retrieve $L(f, \chi, 1)$. We may also apply an arbitrary cohomology class to the universal special value, thus obtaining information about $\Lambda(\chi)$ in the form of "congruences". In good circumstances (§1.5) this leads to congruences satisfied by the algebraic part of $L(f, \chi, 1)$.

To each finite subgroup $B \subseteq \mathrm{Pic}^0(X)_{tor}$ is naturally associated a cohomology class $\varphi_B \in H^1(X; B^\wedge)$ where $B^\wedge$ is the Pontrjagin dual of B. In case B is the cuspidal group, φ_B can be described in terms of periods of weight two Eisenstein series (§1.8). In Chapter 3 we will describe special values associated to the cuspidal group in terms of special values of L-functions attached to Eisenstein series.

The congruence subgroups, $\Gamma \subseteq SL_2(\mathbb{Z})$, and the corresponding modular curves, $X(\Gamma)$ and $Y(\Gamma)$, are introduced in §1.1.

In §1.2 we define the Hecke algebras, $\mathbb{T}$ and $\mathfrak{J}$, and describe our conventions for letting them act on the various homology and cohomology groups attached to Γ.

The cusps of Γ will play an important role in what follows. In §1.3 we describe the action of Galois and of the Hecke operators on them.

The main result of §1.4 states that the groups $H^1(X(\Gamma);\mathbb{Z})$ and $H_1(X(\Gamma);\mathbb{Z})$ are rank 2 $\mathbb{T}$-modules. The section concludes with a discussion of periods of weight two cusp forms.

The concept of a congruence between a cohomology class with values in a finite group and the cohomology class of a cusp form is discussed in §1.5. Proposition 1.5.1 gives simple sufficient conditions for the existence of a cusp form whose cohomology class is congruent to a given cohomology class with values in a finite group.

In §1.6 the universal special value attached to a Dirichlet character χ is defined as a certain 1-cycle $\Lambda(\chi) \in H_1(X(\Gamma);\mathbb{Z}[\chi])$. Proposition 1.6.3 gives the relation between this and the special values, $L(f,\chi,1)$, of the L-functions associated to cusp forms f for Γ. Definition 1.6.4 defines the special value $\Lambda(\varphi,\chi) \in A[\chi]$ associated to an arbitrary cohomology class $\varphi \in H^1(X(\Gamma);A)$ with values in a finite group A.

It is a standard fact that points of finite order on $\mathrm{Pic}^0(X(\Gamma))$ may be identified with cohomology classes on $X(\Gamma)$ with $\mathbb{Q}/\mathbb{Z}$ coefficients. This identification is described in §1.7. We also describe the correspondence between finite subgroups $B \subseteq \mathrm{Pic}^0(X(\Gamma))$ and cohomology classes $\varphi_B \in H^1(X; B^\wedge)$. We give a necessary and sufficient condition that φ_B be congruent to the class of a given cusp form f.

Finally, we examine a distinguished subgroup of $\mathrm{Pic}^0(X)$, namely the cuspidal group $C(\Gamma)$, which we know is finite by the Manin-Drinfeld theorem. Our proof of this fact is essentially theirs, but we give the proof on the level of cohomology in order to emphasize the relation of the cuspidal group to the space of weight two Eisenstein series. A general discussion of periods of Eisenstein series follows. We associate to an arbitrary Eisenstein series, E, a subgroup $C_E \subseteq C(\Gamma)$, and a cohomology class $\varphi_E \in H^1(X(\Gamma); A(E))$.

§1.1. Modular Curves:

Let $\mathcal{H} = \{z \in \mathbb{C} \mid \operatorname{Im}(z) > 0\}$ be the upper half plane and

$$\mathcal{H}^* = \mathcal{H} \cup \mathbb{Q} \cup \{i\infty\} = \mathcal{H} \cup \mathbf{P}^1(\mathbb{Q})$$

be the extended upper half plane obtained by adjoining the cusps, $\mathbf{P}^1(\mathbb{Q})$, and given the usual horocycle topology. The group $GL_2^+(\mathbb{R})$ acts on $\mathcal{H}$ by

$$\alpha = \begin{pmatrix} a & b \\ c & d \end{pmatrix} : z \longmapsto \alpha z = \frac{az+b}{cz+d} \in \mathcal{H} \quad .$$

The center of $GL_2^+(\mathbb{R})$ acts trivially on $\mathcal{H}$. The stabilizer of the point i is the product of the center with the standard maximal compact subgroup $SO_2(\mathbb{R})$. This gives us the well-known description of $\mathcal{H}$ as a symmetric space:

$$\mathcal{H} = PSL_2(\mathbb{R})/PSO_2(\mathbb{R}) \quad .$$

The action of $GL_2^+(\mathbb{Q})$ on $\mathcal{H}$ extends to a continuous action on $\mathcal{H}^*$ which preserves the cusps.

The principal congruence groups are the subgroups $\Gamma(N)$ of $SL_2(\mathbb{Z})$ defined by

$$\Gamma(N) = \{\alpha = \begin{pmatrix} a & b \\ c & d \end{pmatrix} \in SL_2(\mathbb{Z}) \mid \alpha \equiv \pm\begin{pmatrix} 1 & 0 \\ 0 & 1 \end{pmatrix} \pmod N\}$$

where N is a positive integer. The congruence groups are the subgroups, $\Gamma \subseteq GL_2^+(\mathbb{Q})$, which are commensurable with $SL_2(\mathbb{Z})$ and contain a principal congruence group. The smallest integer $N > 0$ for which

$$\Gamma(N) \subseteq \Gamma$$

is called the level of Γ.

We will be mostly interested in the following special congruence groups:

$$\Gamma_0(N) = \{\begin{pmatrix} a & b \\ c & d \end{pmatrix} \in SL_2(\mathbf{Z}) \mid c \equiv 0 \pmod{N}\}$$

$$\Gamma_1(N) = \{\begin{pmatrix} a & b \\ c & d \end{pmatrix} \in \Gamma_0(N) \mid a \equiv d \equiv \pm 1 \pmod{N}\} \quad .$$

If N_1, N_2 are positive integers, let $N = \operatorname{lcm}(N_1, N_2)$ and

$$\Gamma_0(N_1, N_2) = \{\begin{pmatrix} a & b \\ c & d \end{pmatrix} \in \Gamma_0(N_1) \mid b \equiv 0 \pmod{N_2}\} \quad ,$$

$$\Gamma_1(N_1, N_2) = \{\begin{pmatrix} a & b \\ c & d \end{pmatrix} \in \Gamma_0(N_1, N_2) \mid a \equiv d \equiv \pm 1 \pmod{N}\} \quad .$$

We will refer to an intermediate group, Γ, satisfying

$$\Gamma_1(N_1, N_2) \subseteq \Gamma \subseteq \Gamma_0(N_1, N_2)$$

as a group of type (N_1, N_2).

For a congruence group Γ define the modular curves,

$$Y(\Gamma) = \Gamma\backslash\mathcal{H} \quad ,$$

$$X(\Gamma) = \Gamma\backslash\mathcal{H}^* \quad .$$

Write $X(N)$ for $X(\Gamma(N))$. For $i = 0$ or 1 let $X_i(N) = X(\Gamma_i(N))$ and $X_i(N_1, N_2) = X(\Gamma_i(N_1, N_2))$. Similarly for Y.

The modular curve $X(\Gamma)$ may be given the structure of compact Riemann surface. There are finitely many Γ-orbits in $\mathbb{P}^1(\mathbb{Q})$ which represent the finitely many cusps of $X(\Gamma)$. Denote this set by $\underline{\text{cusps}}(\Gamma)$. By removing these cusps we obtain the open Riemann surface $Y(\Gamma)$ which we refer to as the affine part of $X(\Gamma)$.

One of the first remarkable facts about the curves $X(\Gamma)$ is that they may be given the structure of nonsingular projective algebraic curves defined over abelian extensions of $\mathbb{Q}$. In fact the curves $X(\Gamma)$ for Γ of type (N_1, N_2) have models defined over $\mathbb{Q}$ such that the sequence of projections

$$X_1(N_1, N_2) \longrightarrow X(\Gamma) \longrightarrow X_0(N_1, N_2)$$

and projections to lower levels are defined over $\mathbb{Q}$. If Γ is an arbitrary congruence group of level N, then $X(\Gamma)$ may be defined over $\mathbb{Q}(\zeta_N)$. The cusps of $X(\Gamma)$ are defined over $\mathbb{Q}(\zeta_N)$ in this model.

For details see Shimura [42] Chapters 1 and 6.

§1.2. Hecke Operators:

Let N_1, N_2 be positive integers and Γ be a congruence group of type (N_1, N_2). In this section we describe the abstract Hecke algebra $\mathfrak{T}$ and the Hecke algebra $\mathbb{T}$ of Γ. Our basic reference is Shimura [42], especially §§3.3, 3.5, and 7.1-7.3).

For each $g \in GL_2^+(\mathbb{Q})$ let

$$\Gamma(g) = \Gamma \cap g^{-1} \Gamma g \quad .$$

Since $\Gamma(g) \subseteq \Gamma$ there is a natural surjection

$$X(\Gamma(g)) \xrightarrow{\pi(g)} X(\Gamma) \quad .$$

Since $g\Gamma(g)g^{-1} = \Gamma(g^{-1})$, there is a map

$$X(\Gamma(g)) \xrightarrow{g} X(\Gamma(g^{-1})) \quad .$$

Now consider the diagram

$$\begin{array}{ccc} X(\Gamma(g)) & \xrightarrow{g} & X(\Gamma(g^{-1})) \\ {\scriptstyle \pi(g)}\downarrow & & \downarrow{\scriptstyle \pi(g^{-1})} \\ X(\Gamma) & & X(\Gamma) \end{array} \quad .$$

This diagram defines a proper algebraic correspondence, $T(g)$, on $X(\Gamma) \times X(\Gamma)$ by

$$T(g) : X(\Gamma(g)) \xrightarrow{\pi(g) \times (\pi(g^{-1}) \circ g)} X(\Gamma) \times X(\Gamma) \quad .$$

We call T(g) the Hecke correspondence associated to g.

The correspondence T(g) may also be described in terms of the double coset $\Gamma g \Gamma$. Write $\Gamma g \Gamma$ as a disjoint union of right cosets,

$$\Gamma g \Gamma = \bigcup_{i=1}^{k} g_i \Gamma ,$$

and let

$$\pi_\Gamma : \mathcal{H}^* \longrightarrow X(\Gamma)$$

be the natural projection. Then

$$T(g) \cdot \pi_\Gamma(z) = \sum_{i=1}^{k} \pi_\Gamma(g_i z) .$$

Since the Hecke correspondences preserve the cusps of $X(\Gamma)$ they define operators on all of the following homology groups by
$T(g) \mapsto \pi(g^{-1})_* \circ g \circ \pi(g)^*$:

$$H_*(X(\Gamma) ; A) ,$$

$$H_1(Y(\Gamma) ; A) ,$$

$$H_1(X(\Gamma) , \underline{\text{cusps}} (\Gamma) ; A) ,$$

$$H_0(\underline{\text{cusps}} (\Gamma) ; A) ,$$

where A is any abelian group. By duality the Hecke correspondences also operate on the following cohomology groups by $T(g) \mapsto \pi(g)_* \circ g^{-1} \circ \pi(g^{-1})^*$:

$$H^*(X(\Gamma)\,;\,A)\quad,$$

$$H^1(X(\Gamma)\,,\,\underline{\text{cusps}}\,(\Gamma)\,;\,A)\quad,$$

$$H^1(Y(\Gamma)\,;\,A)\quad,$$

$$H^0(\underline{\text{cusps}}\,(\Gamma)\,;\,A)\quad.$$

For each prime ℓ let $\lambda_\ell \in GL_2^+(\mathbb{Q})$ be the matrix

$$\lambda_\ell = \begin{pmatrix} 1 & 0 \\ 0 & \ell \end{pmatrix} \quad .$$

We then write T_ℓ for the Hecke correspondence $T(\lambda_\ell)$. Let $N = \operatorname{lcm}(N_1, N_2)$. For each $m \in (\mathbb{Z}/N\mathbb{Z})^*$ let $\sigma_m \in SL_2(\mathbb{Z})$ such that

$$\sigma_m \equiv \begin{pmatrix} * & 0 \\ 0 & m \end{pmatrix} \pmod N$$

and write $\langle m \rangle$ for $T(\sigma_m)$. For $z \in \mathcal{H}^*$ these correspondence are given by:

$$T_\ell \cdot \pi_\Gamma(z) = \begin{cases} \displaystyle\sum_{k=0}^{\ell-1} \pi_\Gamma\left(\frac{z + N_2 k}{\ell}\right) + \pi_\Gamma(\sigma_\ell(\ell z))\,, & \text{if } \ell \nmid N\ , \\ \displaystyle\sum_{k=0}^{\ell-1} \pi_\Gamma\left(\frac{z + N_2 k}{\ell}\right) & \text{if } \ell \mid N\ . \end{cases}$$

and

$$\langle m \rangle \cdot \pi_\Gamma(z) = \pi_\Gamma(\sigma_m(z)) \quad .$$

Note that when $\ell \mid N$ our operator T_ℓ coincides with the Atkin operator

U_ℓ .

These operators are mutually commutative ([42], Thm. 3.34). Let $\mathfrak{J}$ be the commutative algebra generated by them over $\mathbb{Z}$:

$$\mathfrak{J} = \mathbb{Z}[T_\ell , \langle m \rangle \mid \ell \text{ prime}, m \in (\mathbb{Z}/N\mathbb{Z})^*] \quad .$$

We refer to $\mathfrak{J}$ as the abstract Hecke algebra for Γ . Let $\mathbb{T}$ be the image of the map

$$\mathfrak{J} \longrightarrow \mathrm{End}\,(J(\Gamma)) \quad ,$$

where $J(\Gamma)$ is the Jacobian variety of $X(\Gamma)$.

The algebra $\mathbb{T}$ is a ring of endomorphisms of $J(\Gamma)/\mathbb{Q}$ ([42], Prop. 7.7). By functoriality, $\mathbb{T}$ acts on the following groups

$$H^1(X(\Gamma)\,;A) \qquad (\cong H^1(J(\Gamma)\,;A)) \quad ,$$

$$H_1(X(\Gamma)\,;A) \qquad (\cong H_1(J(\Gamma)\,;A)) \quad ,$$

$$H^0(X(\Gamma)\,;\Omega^1) \qquad (\cong \text{cotangent plane of } J(\Gamma) \text{ at the origin}) \quad .$$

Of course $\mathbb{T}$ also acts on $\mathrm{Pic}^0(X(\Gamma))/\mathbb{Q}$ by duality.

Since $\mathbb{T}$ acts faithfully on $H_1(X(\Gamma)\,;\mathbb{Z})$, $\mathbb{T}$ is a commutative algebra of finite type over $\mathbb{Z}$.

It is important to note that the radical of $\mathbb{T}$ may be nonzero. The nonvanishing of the radical reflects the existence of "old forms" for Γ ([1], [23]).

§1.3. The Cusps:

Let Γ be a congruence group of type (N_1, N_2). Then cusps = cusps(Γ) may be identified with the set

$$\left\{ \begin{pmatrix} x \\ y \end{pmatrix} \in \mathbb{Z}^2 \,\middle|\, (x, y, N_1 N_2) = 1 \right\} / \sim$$

modulo the equivalence relation

$$\begin{pmatrix} x \\ y \end{pmatrix} \sim \begin{pmatrix} x' \\ y' \end{pmatrix} \Longleftrightarrow \begin{pmatrix} x' \\ y' \end{pmatrix} = \begin{pmatrix} ax + by \\ cx + dy \end{pmatrix}$$

for some $\begin{pmatrix} a & b \\ c & d \end{pmatrix} \in \Gamma$. Denote the equivalence class of $\begin{pmatrix} x \\ y \end{pmatrix}$ by $\begin{bmatrix} x \\ y \end{bmatrix}_\Gamma$ or simply $\begin{bmatrix} x \\ y \end{bmatrix}$ if Γ is understood.

For two relatively prime integers x, y write $\{x/y\}_\Gamma \in$ cusps for the cusp on X represented by x/y. Then

$$\{x/y\}_\Gamma \longleftrightarrow \begin{bmatrix} x \\ y \end{bmatrix}$$

gives the identification of cusps with equivalence classes of ordered pairs.

Let $N = \mathrm{lcm}(N_1, N_2)$, then we have the following:

$$(1) \quad \begin{pmatrix} x \\ y \end{pmatrix} \equiv \pm \begin{pmatrix} x' \\ y' \end{pmatrix} \pmod N \Longrightarrow \begin{bmatrix} x \\ y \end{bmatrix} = \begin{bmatrix} x' \\ y' \end{bmatrix};$$

$$(2) \quad \begin{bmatrix} x + N_2 y \\ y \end{bmatrix} = \begin{bmatrix} x \\ y \end{bmatrix} = \begin{bmatrix} x \\ y + N_1 x \end{bmatrix}.$$

The remainder of this section is devoted to describing the action of

Galois and of the Hecke operators on cuspidal divisors.

In order to describe the Galois action we must first specify a model for $X(\Gamma)/\mathbb{Q}$. This is accomplished by giving a function field $\mathfrak{F}(\Gamma)$ for $X(\Gamma)$ which is defined over $\mathbb{Q}$. For this we refer to Shimura ([42], Chapter 6) and let $\mathfrak{F}(\Gamma)$ be the field of modular function for Γ whose Fourier expansions at the infinity cusp have rational coefficients.

THEOREM 1.3.1:

(a) The cusps of X are rational over $\mathbb{Q}(\zeta_N)$, $(\zeta_N = e^{2\pi i/N})$.

(b) For $d \in (\mathbb{Z}/N\mathbb{Z})^*$ let $\tau_d \in \mathrm{Gal}(\mathbb{Q}(\zeta_N)/\mathbb{Q})$ be defined by

$$\tau_d : \zeta_N \longmapsto \zeta_N^d .$$

Then

$$[{x \atop y}]^{\tau_d} = [{x \atop d'y}]$$

where $d' \in \mathbb{Z}$ is chosen so that $dd' \equiv 1 \pmod N$.

<u>Proof:</u> The function field $\mathfrak{F}(\Gamma)$ is contained in the field $\mathfrak{F}_N = \mathbb{Q}(j, f_{\underline{a}} : \underline{a} \in \frac{1}{N}\mathbb{Z}^2/\mathbb{Z}^2 \setminus \{(0,0)\})$ where $f_{\underline{a}}$ denotes the Fricke function (see §6.1 of [42]). The field $\mathfrak{F}_N$ is the field of modular functions for $\Gamma(N)$ with Fourier expansions rational over $\mathbb{Q}(\zeta_N)$. The group $GL_2(\mathbb{Z}/N\mathbb{Z})$ acts on $\mathfrak{F}_N$ by

$$j \mid \gamma = j$$

$$f_{\underline{a}} \mid \gamma = f_{\underline{a}\gamma}$$

for $\gamma \in GL_2(\mathbb{Z}/N\mathbb{Z})$ and $\underline{a} \in \frac{1}{N}\mathbb{Z}^2/\mathbb{Z}^2 \setminus \{\underline{0}\}$. The group $Gal(\mathbb{Q}(\zeta_N)/\mathbb{Q})$ acts on $\mathfrak{F}_N$ through its action on the Fourier coefficients. If we set

$$\tau(d) = \begin{pmatrix} 1 & 0 \\ 0 & d \end{pmatrix} \in GL_2(\mathbb{Z}/N\mathbb{Z})$$

for $d \in (\mathbb{Z}/N\mathbb{Z})^*$, then we have the relation:

$$f^{\tau_d} = f \mid \tau(d)$$

for $f \in \mathfrak{F}_N$.

For a function $f \in \mathfrak{F}_N$ whose Fourier expansion is

$$f(z) = \sum_{n=-m}^{\infty} a_n e^{2\pi i n z/N} , \qquad (m \geq 0)$$

let $\tilde{f}$ be defined by

$$\tilde{f}(q) = \sum_{n=-m}^{\infty} a_n q^n .$$

Let $\gamma \in SL_2(\mathbb{Z})$ be such that $\gamma \cdot (i\infty) = x/y$. A function $f \in \mathfrak{F}(\Gamma)$ is regular at $[{x \atop y}]_\Gamma$ if and only if $\widetilde{f|\gamma}$ has no pole at the origin. If f is regular at $[{x \atop y}]$, then we have

$$f\left(\left[{x \atop y}\right]\right) = \widetilde{f|\gamma}\,(0) .$$

In other words $f\left(\left[{x \atop y}\right]\right)$ is just the constant term of the Fourier expansion of f at x/y.

Now (a) is an immediate consequence of the fact that $f|\gamma \in \mathfrak{F}_N$ and hence has Fourier coefficients in $\mathbb{Q}(\zeta_N)$.

To prove (b), let $Q = \left[{x \atop y}\right]^{\tau_d}$, and let $f \in \mathfrak{F}(\Gamma)$ be arbitrary. Let $\gamma \in SL_2(\mathbb{Z})$ be such that $\gamma \cdot (i\infty) = x/y$ as before. Then

$$f(Q) = f\left(\left[{x \atop y}\right]\right)^{\tau_d} = (\widetilde{f|\gamma}(0))^{\tau_d}$$

$$= \widetilde{f|\gamma\tau(d)}(0) \quad .$$

But $f|\gamma \cdot \tau(d) = f|\tau(d) \cdot \tau(d)^{-1} \cdot \gamma \cdot \tau(d) = f|\tau(d)^{-1} \cdot \gamma \cdot \tau(d) = f|\gamma'$ where γ' is an element of $SL_2(\mathbb{Z})$ whose image in $SL_2(\mathbb{Z}/N\mathbb{Z})$ is $\tau(d)^{-1} \cdot \gamma \cdot \tau(d)$. So we have the congruence

$$\gamma' \equiv \begin{pmatrix} x & * \\ d'y & * \end{pmatrix} \pmod N \quad .$$

Hence

$$f(Q) = \widetilde{f|\gamma'}(0) = f\left(\left[{x \atop d'y}\right]\right) \quad .$$

Since this is true for all $f \in \mathfrak{F}(\Gamma)$, it follows $Q = \left[{x \atop d'y}\right]$. □

Next we compute the action of the Hecke operators on the cuspidal divisors. As a word of warning we point out that though

$\mathrm{Jac}(X(\Gamma)) \cong \mathrm{Pic}^0(X(\Gamma))$ canonically as abelian varieties over $\mathbb{C}$, this isomorphism is generally not an isomorphism of $\mathbb{T}$-modules. The action of $\mathbb{T}$ on $\mathrm{Pic}^0(X(\Gamma))$ is dual to the action on $\mathrm{Jac}(X(\Gamma))$. We will be interested in the action of $\mathbb{T}$ on the cuspidal divisor class group viewed as a subgroup of $\mathrm{Pic}^0(X)$. For this reason we describe the action of the dual correspondences ${}^tT_\ell$, ${}^t\langle \ell \rangle$, $\ell \nmid N$ on the group $\mathrm{Div}(\underline{\mathrm{cusps}})$ of divisors supported on $\underline{\mathrm{cusps}}$.

When we speak of the action of $\mathfrak{I}$ on $\mathrm{Div}(\underline{\mathrm{cusps}})$, we will always mean this (dual) action.

THEOREM 1.3.2: Let ℓ be a prime such that $\ell \nmid N$. Then

(a) for $\left[\begin{smallmatrix} x \\ y \end{smallmatrix}\right] \in \underline{\mathrm{cusps}}$,

$$ {}^t\langle \ell \rangle \cdot \left[\begin{matrix} x \\ y \end{matrix}\right] = \left[\begin{matrix} \ell x \\ \ell' y \end{matrix}\right] $$

where $\ell' \in \mathbb{Z}$ satisfies $\ell \ell' \equiv 1 \pmod N$.

(b) the action of ${}^tT_\ell$ on $\mathrm{Div}(\underline{\mathrm{cusps}})$ is given by

$$ {}^tT_\ell = \ell \cdot {}^t\langle \ell \rangle \cdot \tau_\ell^{-1} + \tau_\ell \quad . $$

Proof: (a) Let $\sigma_\ell = \begin{pmatrix} \ell' & Nb \\ Nc & \ell \end{pmatrix} \in SL_2(\mathbb{Z})$ so that

$$ \sigma_\ell \equiv \begin{pmatrix} \ell' & 0 \\ 0 & \ell \end{pmatrix} \pmod N \quad . $$

Then with the notation of §1.2, for $(x, y) = 1$,

$$ {}^t\langle \ell \rangle \cdot \begin{bmatrix} x \\ y \end{bmatrix} = T\left(\sigma_\ell^{-1}\right) \cdot \begin{bmatrix} x \\ y \end{bmatrix} $$

$$ = T\left(\sigma_\ell^{-1}\right) \cdot \{x/y\}_\Gamma $$

$$ = \left\{ \frac{\ell x - Nby}{-Ncx + \ell' y} \right\}_\Gamma $$

$$ = \begin{bmatrix} \ell x \\ \ell' y \end{bmatrix} . $$

(b) We have ${}^tT_\ell = T(\lambda_\ell^{-1}) = T(\begin{pmatrix} \ell & 0 \\ 0 & 1 \end{pmatrix})$. The double coset decomposition for $\begin{pmatrix} \ell & 0 \\ 0 & 1 \end{pmatrix}$ is

$$ \Gamma \cdot \begin{pmatrix} \ell & 0 \\ 0 & 1 \end{pmatrix} \cdot \Gamma = \bigcup_{k=0}^{\ell-1} \begin{pmatrix} \ell & 0 \\ Nk & 1 \end{pmatrix} \cdot \Gamma \cup \begin{pmatrix} 1 & 0 \\ 0 & \ell \end{pmatrix} \sigma_\ell^{-1} \cdot \Gamma . $$

Let $\begin{bmatrix} x \\ y \end{bmatrix} \in$ <u>cusps</u>. We may assume $(x, y) = 1$, and $\ell \nmid y$. We then have

$$ {}^tT_\ell \cdot \begin{bmatrix} x \\ y \end{bmatrix} = {}^tT_\ell \cdot \{x/y\}_\Gamma $$

$$ = \sum_{k=0}^{\ell-1} \left\{ \frac{\ell x}{Nkx + y} \right\}_\Gamma + \left\{ \frac{\ell x - Nby}{\ell(-Ncx + \ell' y)} \right\}_\Gamma $$

$$ = \sum_{k=0}^{\ell-1} \left\{ \frac{\ell x}{Nkx + y} \right\}_\Gamma + \begin{bmatrix} \ell x \\ y \end{bmatrix} . $$

The greatest common divisor $(\ell x, Nkx + y)$ is either 1 or ℓ, and is ℓ for exactly one choice of k modulo ℓ. Let k_0 be this exceptional value of k. For $k \neq k_0$ we have

$$\left\{\frac{\ell x}{Nkx+y}\right\}_\Gamma = \begin{bmatrix} \ell x \\ y \end{bmatrix},$$

since $(\ell, N) = 1$. For $k = k_0$ we have

$$\left\{\frac{\ell x}{Nk_0x+y}\right\}_\Gamma = \begin{bmatrix} x \\ \ell' y \end{bmatrix} .$$

Hence, $^tT_\ell \cdot [{x \atop y}] = \ell \cdot [{\ell x \atop y}] + [{x \atop \ell' y}]$

$$= (\ell \cdot {}^t\langle \ell \rangle \cdot \tau_\ell^{-1} + \tau_\ell) \cdot [{x \atop y}] \quad . \qquad \square$$

§1.4. $\mathbb{T}$-modules and Periods of Cusp Forms:

Let Γ be of type (N_1, N_2). Let $X = X(\Gamma)$ and $\mathbb{T}$ be the Hecke algebra of Γ as in §1.2. For a field K write $\mathbb{T}_K$ for $\mathbb{T} \otimes_{\mathbb{Z}} K$.

Definition 1.4.1 ([28], II 6.4). A $\mathbb{T}$-module M of finite type is said to have rank r if the following two equivalent conditions are satisfied:

1) $M \otimes_{\mathbb{Z}} \mathbb{Q}$ is a free $\mathbb{T}_{\mathbb{Q}}$-module of rank r.

2) There is a field K of characteristic 0 such that $M \otimes_{\mathbb{Z}} K$ is a free $\mathbb{T}_K$-module of rank r. □

For a $\mathfrak{J}$-module M on which an involution ι acts, let

$$M_+ = \{m \in M | \iota(m) = m\} ,$$
$$M_- = \{m \in M | \iota(m) = -m\} ,$$
$$M^+ = M/(1 - \iota)M ,$$
$$M^- = M/(1 + \iota)M .$$

If 2 acts invertibly on M, then

$$M^+ \cong M_+ , \quad M^- \cong M_- ,$$

and M splits as

$$M \cong M^+ \oplus M^- .$$

We will write $\pm$ to denote either $+$ or $-$.

Let ι be the involution on X induced by the complex conjugation involution $z \mapsto -\bar{z}$ on $\mathcal{H}^*$. Then ι defines involutions

$$\iota^* : H^1(X;A) \longrightarrow H^1(X;A)$$

$$\iota_* : H_1(X;A) \longrightarrow H_1(X;A)$$

for any abelian group A. These involutions commute with the Hecke operators.

The space of weight 2 cusp forms for Γ may be identified with $H^0(X;\Omega^1)$ by

$$f \longleftrightarrow \omega(f) = f(z)\,dz \quad .$$

It follows from ([42], Theorem 3.51) that $H^0(X;\Omega^1)$ is a free $\mathbb{T}_{\mathbb{C}}$-module of rank 1. Since the involution ι^* on $H^1(X;\mathbb{C})$ maps $H^{1,0}$ isomorphically onto $H^{0,1}$, $H^1(X;\mathbb{C})$ is a free rank 2 $\mathbb{T}_{\mathbb{C}}$-module. In fact, the $\mathbb{T}$-homomorphisms $(1 \pm \iota^*)$ map $H^{1,0}$ isomorphically onto $H^1(X;\mathbb{C})_{\pm}$. We have then,

PROPOSITION 1.4.2: The modules $H^1(X;\mathbb{Z})^{\pm}$ and $H_1(X;\mathbb{Z})^{\pm}$ are rank 1 $\mathbb{T}$-modules. □

The statement for the homology groups follows from Poincaré duality.

Since $H^0(X;\Omega^1)$ is a free $\mathbb{T}_{\mathbb{C}}$-module of rank 1, there is a one-to-one correspondence

$$\left\{\begin{matrix}\text{normalized}\\ \text{weight 2}\\ \text{parabolic}\\ \mathbb{T}\text{-eigenforms,}\\ f\end{matrix}\right\} \longleftrightarrow \left\{\begin{matrix}\text{homomorphisms}\\ \\ \mathbb{T} \xrightarrow{h_f} \mathbb{C}\end{matrix}\right\} .$$

For an eigenform, f, the homomorphism h_f satisfies

$$f|\alpha = h_f(\alpha)\cdot f \qquad \text{for} \qquad \alpha \in \mathbb{T} .$$

Let $\mathfrak{P}(f) = \ker(h_f)$ and $\mathfrak{O}(f) = \text{image}(h_f)$. Then $\mathfrak{P}(f)$ is a minimal prime ideal in $\mathbb{T}$ and $\mathfrak{O}(f)$ is the ring generated in $\mathbb{C}$ by the Fourier coefficients of f. Let $K(f)$ be the quotient field of $\mathfrak{O}(f)$.

Let $A_f/\mathbb{Q} \hookrightarrow \mathrm{Pic}^0(X)/\mathbb{Q}$ be the abelian subvariety associated to f by Shimura ([42], Theorem 7.14). Then $\mathfrak{O}(f)$ is identified with a ring of rational endomorphisms of A_f. In fact, the endomorphisms in $\mathbb{T} \hookrightarrow \mathrm{End}(\mathrm{Pic}^0(X)/\mathbb{Q})$ preserve the subvariety $A_f/\mathbb{Q}$ and the homomorphism $h_f : \mathbb{T} \twoheadrightarrow \mathfrak{O}(f) \subseteq \mathrm{End}(A_f/\mathbb{Q})$ is the restriction map.

By passing to the dual abelian varieties we obtain an epimorphism $\mathrm{Jac}(X)/\mathbb{Q} \twoheadrightarrow J_f/\mathbb{Q}$ where $J_f/\mathbb{Q}$ is the variety dual to $A_f/\mathbb{Q}$. Let

$$H = H_1(X;\mathbb{Z}) ,$$

$$H(f) = H_1(J_f;\mathbb{Z}) ,$$

where the notation on the right denotes the singular homology groups of the manifold of $\mathbb{C}$-valued points.

By the definition of A_f the natural map $H_1(A_f;\mathbb{Z}) \hookrightarrow H_1(\mathrm{Pic}^0(X);\mathbb{Z})$ is injective. But $H_1(A_f;\mathbb{Z}) \cong H^1(J_f;\mathbb{Z})$ and $H_1(\mathrm{Pic}^0(X);\mathbb{Z}) \cong H^1(X;\mathbb{Z})$ canonically. Hence by Poincaré duality the natural map

$$\varphi(f) : H \twoheadrightarrow H(f)$$

is surjective.

Since $H \otimes \mathbb{Q}$ is a free rank 2 $\mathbb{T}_{\mathbb{Q}}$-module we have $H(f) \otimes \mathbb{Q}$ is a 2-dimensional $K(f)$-vector space. The complex conjugation involution ι preserves A_f and hence induces an involution on $H(f)$. We have an isomorphism

$$H(f) \otimes \mathbb{Q} \cong (H(f)^+ \otimes \mathbb{Q}) \oplus (H(f)^- \otimes \mathbb{Q}) \quad .$$

The spaces $H(f)^{\pm} \otimes \mathbb{Q}$ are 1-dimensional $K(f)$-vector spaces. Let $\varphi(f)^{\pm}$ be the composition

$$\varphi(f)^{\pm} : H \twoheadrightarrow H(f) \twoheadrightarrow H(f)^{\pm} \quad .$$

Let $\varphi_f \in H^1(X;\mathbb{C})$ be the cohomology class represented by the differential form $\omega(f)$ on X whose pullback to the upper half plane is $f(z)\,dz$. We view φ_f as a $\mathbb{T}$-homomorphism

$$\varphi_f : H \longrightarrow \mathbb{C}$$

where $\mathbb{C}$ is given the structure of $\mathbb{T}$-module inherited from the homomorphism $h_f : \mathbb{T} \to \mathbb{C}$. We will write <u>Periods</u> (f) for the image of φ_f.

From the definition of A_f it follows that φ_f factors through $\varphi(f)$. Hence we have a commutative diagram

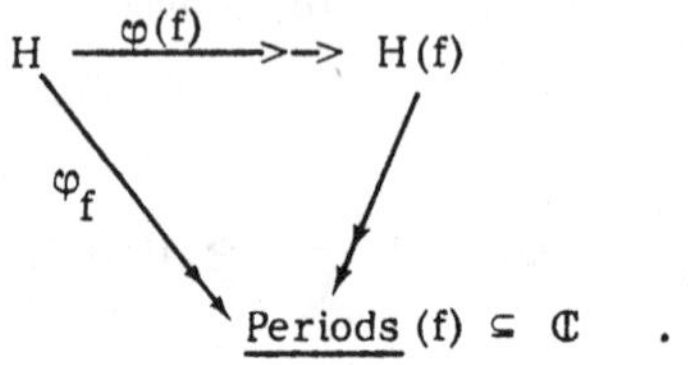

Let $\gamma_{\pm}$ be the generators of the free rank 1 $\mathbb{T}_{\mathbb{Q}}$-modules $H_{\pm} \otimes \mathbb{Q}$. Then each $\gamma \in H \otimes \mathbb{Q}$ may be expressed uniquely in the form

$$\gamma = \alpha \cdot \gamma_+ + \beta \cdot \gamma_-$$

with $\alpha, \beta \in \mathbb{T}_{\mathbb{Q}}$. If we extend φ_f by $\mathbb{Q}$-linearity to a map $H \otimes \mathbb{Q} \to \mathbb{C}$, and let

$$\Omega_f^{\pm} = \varphi_f(\gamma_{\pm})$$

then

$$\varphi_f(\gamma) = h_f(\alpha)\,\Omega_f^{+} + h_f(\beta)\,\Omega_f^{-} \quad .$$

Define $\varphi_f^{\pm} \in H^1(X;\mathbb{C})^{\pm}$ by

$$\varphi_f^{\pm} = \frac{(1 \pm \iota^*)}{2} \cdot \varphi_f \quad .$$

Then for $\gamma = \alpha \cdot \gamma_+ + \beta \cdot \gamma_-$ as above we have

$$\varphi_f^+(\gamma) = h_f(\alpha) \cdot \Omega_f^+ ,$$

$$\varphi_f^-(\gamma) = h_f(\beta) \cdot \Omega_f^- .$$

Let <u>Periods</u> $(f)^\pm = \varphi_f^\pm(H)$. Then $\mathbb{Q} \cdot$ <u>Periods</u> $(f)^\pm = K(f) \cdot \Omega_f^\pm$.

The maps $\varphi_f^\pm$ factor through $\varphi(f)^\pm$

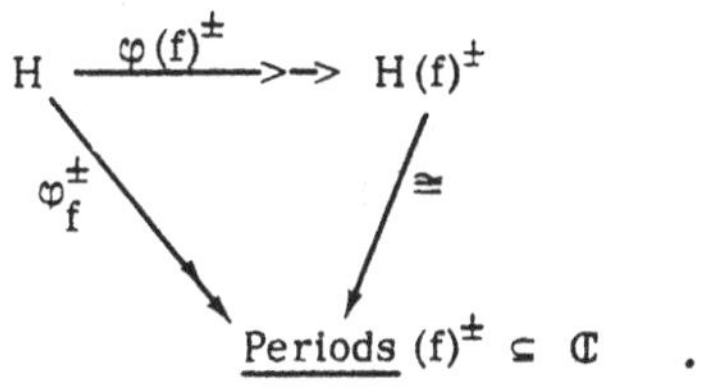

This factorization identifies $H(f)^\pm$ with <u>Periods</u> $(f)^\pm$.

§1.5. Congruences:

We keep the notation of §1.4. Let A be a finite $\mathbb{T}$-module and $\varphi : H \twoheadrightarrow A$ be a surjective $\mathbb{T}$-homomorphism. We will view a commutative diagram

$$\begin{array}{ccc} H & \xrightarrow{\quad\varphi\quad} & A \\ & {\scriptstyle \varphi(f)^{\pm}} \searrow \quad \circ \quad \nearrow & \\ & H(f)^{\pm} & \end{array}$$

as expressing a congruence between φ and $\varphi(f)^{\pm}$ modulo the kernel of φ. In later sections we give explicit constructions of many such homomorphisms φ. It would be interesting to know when the existence of such φ expresses a congruence satisfied by the algebraic part of the periods of a cusp form f. The next proposition is a weak result in this direction.

For a prime ideal $P \subseteq \mathbb{T}$ and a $\mathbb{T}$-module M, write M_P for the completion of M at P.

PROPOSITION 1.5.1: Let A be a finite $\mathbb{T}$-module and $\varphi : H^{\pm} \to A$ a surjective $\mathbb{T}$-homomorphism. Let $P \subseteq \mathbb{T}$ be a maximal ideal for which $H^{\pm}_P$ is a free rank 1 $\mathbb{T}_P$-module. Suppose one of the following two conditions is satisfied:

(i) $A_P \cong \mathbb{T}/P$;

(ii) $\mathbb{T}_P$ is an integral domain.

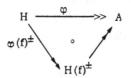

Then for each parabolic eigenform f for which $\mathfrak{P}(f) \subseteq P$, there is a $\mathbb{T}$-homomorphism j which makes the diagram

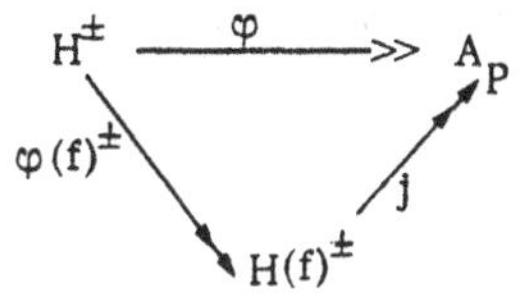

commutative.

Proof: Let f be an eigenform for which $\mathfrak{P}(f) \subseteq P$. The local freeness of $H^{\pm}$ implies

$$H(f)^{\pm}_{P} \cong H^{\pm}_{P}/\mathfrak{P}(f)\, H^{\pm}_{P} \quad .$$

In case (i), let j be the composition of the maps

$$H(f)^{\pm} \longrightarrow H(f)^{\pm}_{P} \longrightarrow H^{\pm}_{P}/PH^{\pm}_{P} \xrightarrow[\sim]{\varphi} A_P$$

In case (ii), $\mathfrak{P}(f)$ is the kernel of localization at P. Hence the map

$$H^{\pm}_{P} \longrightarrow H(f)^{\pm}_{P}$$

is an isomorphism. Let j be the composition

$$H(f)^{\pm} \longrightarrow H(f)^{\pm}_{P} \longrightarrow H^{\pm}_{P} \longrightarrow A_P \quad . \qquad \square$$

Remark 1: In Chapter 3 we will take P to be an Eisenstein prime.

Remark 2: The condition that $\mathbb{T}_P$ is an integral domain is equivalent to saying that P contains precisely one minimal prime ideal $\mathfrak{P}$ and that $\mathbb{T}$ is reduced at $\mathfrak{P}$ (i.e. $\mathrm{rad}(\mathbb{T}_{\mathfrak{P}}) = 0$). The condition that $\mathbb{T}$ be reduced at $\mathfrak{P}$ is satisfied if $\mathfrak{P} = \mathfrak{P}(f)$ for a new form f. This is equivalent to the multiplicity one theorem for new forms.

§1. 6. The Universal Special Values:

Let Γ be a congruence group of type (N_1, N_2), and $X = X(\Gamma)$.

Let

$$f(z) = \sum_{n=1}^{\infty} a_n q_{N_2}^n , \quad q_{N_2} = e^{2\pi i z/N_2}$$

be a normalized weight two cusp form for Γ which is an eigenfunction for $\mathbb{T}$. For a primitive Dirichlet character $\chi \neq 1$ of conductor m prime to $N = \operatorname{lcm}(N_1, N_2)$, let

$$f_\chi(z) = \bar{\chi}(N_2) \sum_{n=1}^{\infty} a_n \chi(n) q_{N_2}^n .$$

The L-function of f twisted by χ is the Mellin transform of f_χ:

$$L(f, \chi, s) = -2\pi i \int_0^{i\infty} f_\chi(z) \operatorname{Im}(z)^{s-1} dz \qquad (s \in \mathbb{C})$$

$$= N_2^s \cdot \bar{\chi}(N_2) \cdot \sum_{n=1}^{\infty} a_n \chi(n) n^{-s} \qquad (\operatorname{Re}(s) > 3/2) .$$

We are interested in the "special value" of $L(f, \chi, s)$ at $s = 1$.

For a ring $R \subseteq \mathbb{C}$, let $R[\chi]$ be the ring in $\mathbb{C}$ generated over R by the values of χ.

Following Mazur ([26], §6) we define the universal special value associated to χ to be a relative homology class

$$\Lambda(\chi) \in H_1(X, \underline{\text{cusps}}; \mathbb{Z}[\chi])$$

as follows.

Definition 1. 6. 1:

(a) For two cusps $x, y \in \mathbb{P}^1(\mathbb{Q}) \subseteq \mathcal{H}^*$ let

$$\{x, y\}_\Gamma \in H_1(X, \underline{\text{cusps}}; \mathbb{Z})$$

be the relative homology class represented by the projection to X of the oriented geodesic path joining x to y.

(b) For a primitive Dirichlet character $\chi \neq 1$ of conductor m prime to N, let

$$\Lambda(\chi) = \sum_{a=0}^{m-1} \bar{\chi}(aN_2) \cdot \left\{ i\infty, \frac{aN_2}{m} \right\}_\Gamma \in H_1(X, \underline{\text{cusps}}; \mathbb{Z}[\chi]) \quad .$$

Remark 1. 6. 2: There is a natural inclusion of the absolute homology of X in the relative homology:

$$H_1(X; \mathbb{Z}) \hookrightarrow H_1(X, \underline{\text{cusps}}; \mathbb{Z}) \quad .$$

The special value $\Lambda(\chi)$ actually represents an absolute homology class. To see this, consider the long exact relative homology sequence

$$0 \longrightarrow H_1(X; \mathbb{Z}[\chi]) \longrightarrow H_1(X, \underline{\text{cusps}}; \mathbb{Z}[\chi]) \xrightarrow{\partial} H_0(\underline{\text{cusps}}; \mathbb{Z}[\chi]) .$$

We need only show $\partial(\Lambda(\chi)) = 0$. We have

$$\partial(\Lambda(\chi)) = \sum_{a=0}^{m-1} \bar{\chi}(aN_2) \cdot \left(\left\{ \frac{aN_2}{m} \right\}_\Gamma - \{i\infty\}_\Gamma \right) \quad .$$

But since $(m, N) = 1$, $\left\{ \frac{aN_2}{m} \right\}_\Gamma = \left\{ \frac{N_2}{m} \right\}_\Gamma$ for all $a \in \mathbb{Z}$ prime to m (see §1. 3). □

Definition 1.6.1 is motivated by the following proposition due to Birch [2].

PROPOSITION 1.6.3: For a primitive Dirichlet character χ of conductor $m > 1$, m prime to N and a cusp form f for Γ,

$$\tau(\bar{\chi})\, L(f, \chi, 1) = -2\pi i \sum_{a=0}^{m-1} \bar{\chi}(aN_2) \int_{\frac{aN_2}{m}}^{i\infty} f(z)\, dz \quad ,$$

where

$$\tau(\bar{\chi}) = \sum_{a=0}^{m-1} \bar{\chi}(a)\, e^{2\pi i a/m}$$

is the usual Gauss sum.

Proof: For $k \in \mathbf{Z}$, let

$$\tau(\bar{\chi}, k) = \sum_{a=0}^{m-1} \bar{\chi}(a)\, e^{2\pi i a k/m} \quad .$$

Then
$$\tau(\bar{\chi}, k) = \chi(k) \cdot \tau(\bar{\chi}) \quad .$$

Suppose $f(z) = \sum_{n=1}^{\infty} a_n q_{N_2}^n$, then

$$\begin{aligned}
\tau(\bar{\chi}) \cdot f_\chi(z) &= \bar{\chi}(N_2) \sum_{n=1}^{\infty} a_n \cdot \tau(\bar{\chi}, n)\, q_{N_2}^n \\
&= \sum_{a=0}^{m-1} \bar{\chi}(aN_2) \sum_{n=1}^{\infty} a_n\, e^{2\pi i a n/m}\, e^{2\pi i n z/N_2} \\
&= \sum_{a=0}^{m-1} \bar{\chi}(aN_2) \sum_{n=1}^{\infty} a_n\, e^{2\pi i n\left(z + \frac{aN_2}{m}\right)/N_2} \\
&= \sum_{a=0}^{m-1} \bar{\chi}(aN_2) \cdot f\left(z + \frac{aN_2}{m}\right) \quad .
\end{aligned}$$

We then have

$$\tau(\overline{\chi}) \cdot L(f, \chi, 1) = -2\pi i \int_0^{i\infty} f_\chi(z)\, dz$$

$$= -2\pi i \sum_{a=0}^{m-1} \overline{\chi}(aN_2) \int_{\frac{aN_2}{m}}^{i\infty} f(z)\, dz \qquad \square$$

Let A be an arbitrary abelian group and let

$$\varphi \in H^1(X; A) \quad .$$

Let χ be a primitive Dirichlet character of conductor m prime to N, and let $A \otimes_{\mathbb{Z}} \mathbb{Z}[\chi] \twoheadrightarrow A[\chi]$ be a surjective $\mathbb{Z}[\chi]$-homomorphism onto a $\mathbb{Z}[\chi]$-module $A[\chi]$.

Definition 1.6.4: The special value associated to the pair φ, χ is

$$\Lambda(\varphi, \chi) \overset{\text{dfn}}{=} \varphi \cap \Lambda(\chi) \in A[\chi] \quad . \qquad \square$$

In particular, if $\varphi_f \in H^1(X; \mathbb{C})$ is the class associated to the differential form f(z) dz, then the last proposition shows

$$2\pi i\, \Lambda(\varphi_f, \chi) = \tau(\overline{\chi}) \cdot L(f, \chi, 1) \quad .$$

By the discussion of §1.4 there are periods $\Omega_f^{\pm}$ such that

$$\varphi_f^{\pm} : H^{\pm} \longrightarrow K(f) \cdot \Omega_f^{\pm} \quad .$$

If $\operatorname{sgn} \chi = \chi(-1)$, we have

$$\Lambda(\varphi_f, \chi) = \Lambda(\varphi_f^{\mathrm{sgn}\,\chi}, \chi) \in K(f)[\chi] \cdot \Omega_f^{\mathrm{sgn}\,\chi} \quad .$$

We will refer to $\Lambda(\varphi_f, \chi)/\Omega_f^{\mathrm{sgn}\,\chi} \in K(f)[\chi]$ as the algebraic part of $L(f, \chi, 1)$. Of course this depends on the choice of the periods $\Omega_f^{\pm}$.

In subsequent sections we will construct explicit cohomology classes φ with values in finite $\mathbb{T}$-modules. By calculating the associated special values we obtain congruences satisfied by the universal special value $\Lambda(\chi)$. In case the conditions of Proposition 1.5.1 are satisfied we also obtain congruences satisfied by the algebraic part of $L(f, \chi, 1)$ for appropriate cusp forms f.

§1.7. Points of finite order in $\mathrm{Pic}^0(X(\Gamma))$:

Let Γ be a congruence group. Let μ_n be the group of n-th roots of 1 in $\mathbb{C}^*$. Using the exact sequence of sheaves on $X = X(\Gamma)$,

$$0 \longrightarrow \mu_n \longrightarrow \mathcal{O}^* \xrightarrow{\ n\ } \mathcal{O}^* \longrightarrow 0$$

we may identify $H^1(X;\mu_n)$ with $\mathrm{Pic}^0(X)[n]$, the group of elements of order n in the Picard group of X.

The exponential map gives an isomorphism

$$\exp : \frac{1}{n}\mathbb{Z}/\mathbb{Z} \xrightarrow{\ \sim\ } \mu_n$$

of abelian groups. For $x \in \mathrm{Pic}^0(X)[n]$, let

$$\varphi_x \in H^1(X;\frac{1}{n}\mathbb{Z}/\mathbb{Z})$$

be such that $\exp(\varphi_x) \in H^1(X;\mu_n)$ is the class corresponding to x under the above identification.

PROPOSITION 1.7: Let $x \in \mathrm{Pic}^0(X)[n]$ and D_x a divisor on X representing x. Let g be a meromorphic function on X with

$$\mathrm{div}(g) = n \cdot D_x \quad .$$

Then for $\gamma \in H_1(X;\mathbb{Z})$

$$\varphi_x \cap \gamma \equiv \frac{1}{2\pi i}\int_\gamma \frac{dg}{ng} \quad (\mathrm{mod}\ \mathbb{Z}) \quad .$$

Proof: Let $\Omega_{\mathbb{Z}}$ be the sheaf of differentials of the third kind on X having integral residues. Let

$$\delta : H^0(X;\Omega_{\mathbb{Z}}) \longrightarrow H^1(X;\mathbb{C}^*)$$

be the coboundary map arising from the exact sequence

$$0 \longrightarrow \mathbb{C}^* \longrightarrow \mathfrak{m}^* \xrightarrow{d\log} \Omega_{\mathbb{Z}} \longrightarrow 0 \quad .$$

Then for $\omega \in H^0(X;\Omega_{\mathbb{Z}})$ and $\gamma \in H_1(X;\mathbb{Z})$

$$\delta(\omega) \cap \gamma = \exp\left(\frac{1}{2\pi i}\int_{\gamma} \omega\right) ,$$

where the right-hand side is well defined because of the integrality of the residues of ω.

On the other hand, if g is as in the proposition, and

$$\omega = \frac{dg}{ng} ,$$

$\delta(\omega)$ is represented by the Čech cocycle

$$\left\{(g|U_\alpha)^{1/n} \cdot (g|U_\beta)^{-1/n}\right\}_{\alpha,\beta} \in \check{Z}^1(\mathfrak{U};X;\mathcal{O}^*)$$

for a sufficiently fine open cover $\mathfrak{U} = \{U_\alpha\}_\alpha$ of X, and arbitrarily chosen n-th roots. But this cocycle represents the class in $H^1(X;\mathcal{O}^*)$ associated to the divisor D_x. Hence

$$\delta(\omega) = \exp(\varphi_x) \quad .$$

□

We will write $G^{\wedge}$ for the Pontrjagin dual of a locally compact abelian group G. There is a canonical isomorphism $\mathrm{Pic}^0(X)^{\wedge} \cong H_1(X;\mathbb{Z})$.

Let $B \subseteq \mathrm{Pic}^0(X)_{\mathrm{tor}}$ be a finite subgroup, and let $A = B^{\wedge}$. By Pontrjagin duality we obtain a homomorphism

$$\varphi_B : H_1(X;\mathbb{Z}) \twoheadrightarrow A$$

which we may view as an element of $H^1(X;A)$. We call φ_B the cohomology class associated to B.

Let $f \in \mathcal{S}_2(\Gamma)$ be a $\mathbb{T}$-eigenform and suppose B is a finite subgroup of $A_{f,\mathrm{tor}}$. Then φ_B is "congruent" to φ_f in the sense of §1.5. This follows immediately by passing to the Pontrjagin dual of the commutative diagram

$$\begin{array}{ccc} \mathrm{Pic}^0(X) & \hookleftarrow & B \\ & \nwarrow & \downarrow \\ & & A_f \end{array} .$$

In the next section we will use Eisenstein series to describe certain finite subgroups of $\mathrm{Pic}^0(X)$ and their associated cohomology classes.

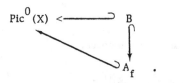

§1.8. Eisenstein Series and the Cuspidal Group:

As before let Γ be of type (N_1, N_2) and $X = X(\Gamma)$. Contained in $\mathrm{Pic}^0(X)$ there is a distinguished subgroup generated by the degree 0 divisor classes supported on the cusps. This group is called the cuspidal group. We denote it by the letter $C = C(\Gamma)$. In this section we prove the Manin-Drinfeld theorem which states that C is finite. We also show how to associate to an arbitrary Eisenstein series E for Γ a subgroup $C_E \subseteq C$ of the cuspidal group.

Let $Y = Y(\Gamma)$ be the affine part of X. Let $\mathfrak{M}_2(\Gamma)$ be the space of weight 2 modular forms for Γ. There is a natural injection $\omega : \mathfrak{M}_2(\Gamma) \hookrightarrow H^1(Y; \mathbb{C})$ which sends a modular form f to the cohomology class represented by the differential form $\omega(f) = f(z)\,dz$. In fact, ω is a homomorphism of $\mathfrak{J}$-modules.

The space $\mathfrak{M}_2(\Gamma)$ splits into the product of the space of cusp forms $\mathcal{S}_2(\Gamma)$ and the space of Eisenstein series $\mathcal{E}_2(\Gamma)$. Then ω maps $\mathcal{S}_2(\Gamma)$ isomorphically onto the space of holomorphic 1-forms on X, and maps $\mathcal{E}_2(\Gamma)$ onto a space of differentials of the third kind on X with simple poles only at the cusps.

For an abelian group A define $\mathrm{Div}^0(\underline{\text{cusps}}; A) \overset{\text{dfn}}{=} \mathrm{Div}^0(\underline{\text{cusps}}) \otimes A$. For $E \in \mathcal{E}_2(\Gamma)$ let $\delta(E) = 2\pi i \sum_{x \in \underline{\text{cusps}}} \mathrm{res}_x(\omega(E)) \cdot \{x\} \in \mathrm{Div}^0(\underline{\text{cusps}}; \mathbb{C})$.

This divisor may also be described as follows. For $x \in \underline{\text{cusps}}$ let $\mathrm{ind}_\Gamma(x)$

denote the ramification index of x over $X(1)$, and let $r_E(x)$ be the constant term of the Fourier expansion of E at x. Then

$$\delta(E) = \sum_{x \in \underline{\text{cusps}}} \text{ind}_\Gamma(x) \cdot r_E(x) \cdot \{x\} .$$

The map $\delta : \mathcal{E}_2(\Gamma) \longrightarrow \text{Div}^0(\underline{\text{cusps}}; \mathbb{C})$ is an isomorphism ([39], Thm. 8, pg. 171).

The long exact homology sequence of the pair $(X, \underline{\text{cusps}})$ gives the exact sequence

$$0 \longrightarrow H_1(X;A) \longrightarrow H_1(X, \underline{\text{cusps}}; A) \longrightarrow \widetilde{H}_0(\underline{\text{cusps}}; A) \longrightarrow 0$$

where $\widetilde{H}_0$ denotes reduced homology. By Lefschetz duality we obtain the following exact sequence of $\mathfrak{J}$-modules:

$$0 \longrightarrow H^1(X;A) \longrightarrow H^1(Y;A) \longrightarrow \widetilde{H}^0(\underline{\text{cusps}}; A) \longrightarrow 0 .$$

There is a natural isomorphism $\widetilde{H}^0(\underline{\text{cusps}}; A) \cong \text{Div}^0(\underline{\text{cusps}}; A)$. With the convention of §1.3 concerning the action of $\mathfrak{J}$ on $\text{Div}^0(\underline{\text{cusps}})$ this is a $\mathfrak{J}$-isomorphism. So we have the following exact sequence of $\mathfrak{J}$-modules:

$$(1.8.1) \quad 0 \longrightarrow H^1(X;A) \xrightarrow{i} H^1(Y;A) \xrightarrow{r} \text{Div}^0(\text{cusps}; A) \longrightarrow 0 .$$

Let $s = \omega \circ \delta^{-1} : \text{Div}^0(\underline{\text{cusps}}; \mathbb{C}) \longrightarrow H^1(Y; \mathbb{C})$. Then s is a $\mathfrak{J}$-homomorphism and $r \circ s = \text{id}$. Hence, putting $A = \mathbb{C}$ in (1.8.1), s gives us a splitting

$$H^1(Y;\mathbb{C}) = H^1(X;\mathbb{C}) \oplus \omega(\mathcal{E}_2(\Gamma)) \quad .$$

In the next proposition we show that this splitting respects the $\mathbb{Q}$-structure.

Let $R \subseteq \mathbb{C}$ be an arbitrary subgroup of $\mathbb{C}$, and define

$$\mathcal{E}_2(\Gamma;R) \overset{\text{dfn}}{=} \delta^{-1}(\mathrm{Div}^0(\underline{\text{cusps}};R))$$
$$\subseteq \mathcal{E}_2(\Gamma) \quad .$$

THEOREM 1.8.2: $\omega(\mathcal{E}_2(\Gamma;\mathbb{Q})) \subseteq H^1(Y;\mathbb{Q})$.

Hence $H^1(Y;\mathbb{Q}) = H^1(X;\mathbb{Q}) \oplus \omega(\mathcal{E}_2(\Gamma;\mathbb{Q}))$.

Remark: It is possible to prove this by producing explicit formulae for the periods of $\omega(E)$ (and noting that they are rational). We will do this in the next chapter. The proof we give here uses the Hecke operators and is due (in essence) to Manin ([25], Corollary 3.6) and Drinfeld [6].

Proof of Theorem 1.8.2: We will show that the map $s : \mathrm{Div}^0(\underline{\text{cusps}};\mathbb{C}) \longrightarrow H^1(Y;\mathbb{C})$ is defined over $\mathbb{Q}$ which will prove the theorem.

Define a projection $\pi : H^1(Y;\mathbb{C}) \longrightarrow H^1(X;\mathbb{C})$ by the identity

$$i \circ \pi = id - s \quad r \quad .$$

Then π is a $\mathfrak{J}$-homomorphism.

Let $p \equiv 1 \pmod{N_1 N_2}$ be a prime and set $\eta_p = T_p - p - 1 \in \mathfrak{J}$. The operator on $H^1(X;\mathbb{C})$ defined by T_p has eigenvalues with absolute value less than $p+1$ ([42], Theorem 7.12). Therefore η_p acts invertibly on $H^1(X;\mathbb{C})$. Let φ_p be the inverse map.

By Theorem 1.3.2(b), η_p annihilates $\mathrm{Div}^0(\underline{\text{cusps}};\mathbb{C})$ and hence determines a map

$$\bar{\eta}_p : H^1(Y;\mathbb{C}) \longrightarrow H^1(X;\mathbb{C})$$

satisfying $i \circ \bar{\eta}_p = \eta_p$. We then have $\eta_p \circ \pi = \pi \circ \eta_p = \bar{\eta}_p$ from which we conclude

$$\pi = \varphi_p \circ \bar{\eta}_p \quad .$$

Since the maps φ_p and $\bar{\eta}_p$ are defined over $\mathbb{Q}$, so is π, and finally so is s. □

Define a map $x : \mathcal{E}_2(\Gamma;\mathbb{Z}) \longrightarrow\!\!\!\!\!\rightarrow C(\Gamma)$, $x : E \mapsto x(E)$ by the composition

$$\mathcal{E}_2(\Gamma;\mathbb{Z}) \xrightarrow[\sim]{\delta} \mathrm{Div}^0(\underline{\text{cusps}}) \longrightarrow\!\!\!\!\!\rightarrow C(\Gamma) \quad .$$

COROLLARY 1.8.3: (a) $C(\Gamma)$ is finite.

(b) For $E \in \mathcal{E}_2(\Gamma)$ and $\gamma \in H_1(X;\mathbb{Z})$

$$\varphi_{x(E)} \cap \gamma \equiv \int_{\tilde{\gamma}} \omega(E) \qquad (\mathrm{mod}\ \mathbb{Z})$$

where $\tilde{\gamma} \in H_1(Y;\mathbb{Z})$ is any element whose image is γ under the natural map

$$H_1(Y;\mathbb{Z}) \longrightarrow\!\!\!\to H_1(X;\mathbb{Z}) \quad .$$

Proof: To prove (a) it suffices to show that $x(E) \in C(\Gamma)$ has finite order for each $E \in \mathcal{E}_2(\Gamma;\mathbb{Z})$. By the theorem $\omega(E)$ has rational periods on Y. Since $H_1(Y;\mathbb{Z})$ is finitely generated, there is an $n \in \mathbb{Z}$ such that $n \cdot \omega(E) \in H^1(Y;\mathbb{Z})$.

Choose an arbitrary point $z_0 \in \mathcal{H}$ and view $\omega(E)$ as a differential form on $\mathcal{H}$ by pullback. Define a function g on $\mathcal{H}$ by

$$g(z) = \exp\left(\int_{z_0}^{z} n \cdot \omega(E)\right) \quad .$$

Then g is invariant for Γ and defines a meromorphic function on X with

$$\operatorname{div}(g) = n \cdot \delta(E) \quad .$$

Since $\delta(E)$ represents the divisor class $x(E)$ we have $n \cdot x(E) = 0$.

To prove (b) we note $\omega(E) = \frac{1}{2\pi i} \cdot \frac{dg}{ng}$ and apply Proposition 1.7.

□

Let $A, B \subseteq \mathbb{C}$ be two subgroups of $\mathbb{C}$ and $\phi : A \to B$ be a homomorphism. Then ϕ defines a map $\phi : \mathrm{Div}^0(\underline{\text{cusps}};A) \to \mathrm{Div}^0(\underline{\text{cusps}};B)$ and hence also $\phi : \mathcal{E}_2(\Gamma;A) \to \mathcal{E}_2(\Gamma;B)$. By the theorem $\omega(\mathcal{E}_2(\Gamma;A)) \subseteq H^1(Y;\mathbb{Q}\cdot A)$. We may extend ϕ to a $\mathbb{Q}$-linear map $\phi : \mathbb{Q}\cdot A \to \mathbb{Q}\cdot B$ and then ϕ also defines a map $\phi : H^1(Y;\mathbb{Q}\cdot A) \to H^1(Y;\mathbb{Q}\cdot B)$.

COROLLARY 1.8.4: For $\phi : A \to B$, and $E \in \mathcal{E}_2(\Gamma; A)$ we have

$$\omega(\phi \cdot E) = \phi \cdot \omega(E) \quad .$$

Proof: E can be expressed as a linear combination $E = \sum_{i=1}^{n} a_i \cdot E_i$ with $E_i \in \mathcal{E}_2(\Gamma; \mathbb{Z})$. By the theorem, $\phi \cdot E = \sum_{i=1}^{n} \phi(a_i) \cdot E_i$. Hence

$$\omega(\phi \cdot E) = \sum_{i=1}^{n} \phi(a_i)\, \omega(E_i) = \phi\left(\sum_{i=1}^{n} a_i\, \omega(E_i)\right)$$

$$= \phi \cdot \omega(E) \quad . \qquad \square$$

Now let K be a number field and $E \in \mathcal{E}_2(\Gamma; K)$. Let $R(E) \subseteq K$ be the $\mathbb{Z}$-submodule of K generated by the coefficients of $\delta(E)$. Integration of $\omega(E)$ gives a map $H_1(Y; \mathbb{Z}) \to K$. Let Periods (E) denote the image of this map.

Let Par $\subseteq H_1(Y; \mathbb{Z})$ be the parabolic subgroup. That is, Par is the group generated by the homology classes represented by small loops around the deleted cusps. Then we have a diagram of exact rows and surjective columns

$$\begin{array}{ccccccccc} 0 & \longrightarrow & \text{Par} & \longrightarrow & H_1(Y; \mathbb{Z}) & \longrightarrow & H_1(X; \mathbb{Z}) & \longrightarrow & 0 \\ & & \Big\downarrow {\scriptstyle \int \omega(E)} & & \Big\downarrow {\scriptstyle \int \omega(E)} & & \Big\downarrow {\scriptstyle \varphi_E} & & \\ 0 & \longrightarrow & R(E) & \longrightarrow & \underline{\text{Periods}}\,(E) & \longrightarrow & A(E) & \longrightarrow & 0 \end{array}$$

where $A(E) \overset{\text{dfn}}{=} \underline{\text{Periods}}(E)/_{R(E)}$, and φ_E is the unique map which makes the diagram commute. We may view φ_E as a cohomology class

$$\varphi_E \in H^1(X; A(E)) \quad .$$

By Theorem 1.8.2 $\underline{\text{Periods}}(E) \subseteq \mathbb{Q} \cdot R(E)$. Hence $A(E)$ is finite.

Let $R(E)^* = \{\phi \in \mathrm{Hom}_{\mathbb{Q}}(\mathbb{Q} \cdot R(E); \mathbb{Q}) \mid \phi(R(E)) \subseteq \mathbb{Z}\}$. Then $R(E)^* \cong \mathrm{Hom}(R(E); \mathbb{Z})$.

Definition 1.8.5: The subgroup $C_E \subseteq C(\Gamma)$ associated to E is the image of the composition

$$\begin{array}{ccccc} R(E)^* & \hookrightarrow & \mathrm{Div}^0(\underline{\text{cusps}}) & \twoheadrightarrow & C(\Gamma) \\ \phi & \longmapsto & \phi(\delta(E)) & . & \end{array}$$

□

The following proposition shows that $\varphi_E \in H^1(X; A(E))$ is the cohomology class associated to C_E (§1.7). By Pontrjagin duality we obtain a map $\hat{\varphi}_E : A(E)^\wedge \to \mathrm{Pic}^0(X)_{tor}$.

PROPOSITION 1.8.6: The image of $\hat{\varphi}_E$ is the group C_E.

Proof: There is a natural surjection $R(E)^* \twoheadrightarrow A(E)^\wedge$. To prove the proposition it suffices to show that the following diagram is commutative.

$$\begin{array}{ccc} R(E)^* & \longrightarrow\!\!\!\!\twoheadrightarrow & A(E)^\wedge \\ \cap\downarrow & \cdot & \cap\downarrow \\ \mathrm{Div}^0(\underline{\mathrm{cusps}}) & \twoheadrightarrow & C(\Gamma) \subseteq \mathrm{Pic}^0(X) \cong H_1(X, \mathbb{Z})^\wedge \end{array}$$

Let $\phi \in R(E)^*$. The image of ϕ in $\mathrm{Div}^0(\underline{\mathrm{cusps}})$ is $\delta(\phi E)$ and the image of this in $C(\Gamma)$ is $x(\phi E)$. If $\psi : A(E) \to \mathbb{Q}/\mathbb{Z}$ is the image of ϕ in $A(E)^\wedge$ then we have, for $\gamma \in H_1(X;\mathbb{Z})$:

$$\begin{aligned} \varphi_{x(\phi E)} \cap \gamma &\equiv \int_{\tilde{\gamma}} \omega(\phi E) \pmod{\mathbb{Z}} && \text{(Cor. 1.8.3 (b))} \\ &= \phi\left(\int_{\tilde{\gamma}} \omega(E)\right) && \text{(Cor. 1.8.4)} \\ &\equiv \psi \circ \varphi_E(\gamma) \quad . && \square \end{aligned}$$

Let $\underline{\mathrm{Periods}}(E)^*$ be the group of $\phi \in R(E)^*$ for which $\phi(\underline{\mathrm{Periods}}(E)) \subseteq \mathbb{Z}$. By Corollary 1.8.3 we have the following.

COROLLARY 1.8.7: The sequence

$$0 \longrightarrow \underline{\mathrm{Periods}}(E)^* \longrightarrow R(E)^* \longrightarrow C_E \longrightarrow 0$$

is exact. $\square$

Chapter 2. Periods of Modular Forms

In this chapter we develop the tools needed to describe the subgroup of $H^1(X(\Gamma);\mathbb{Q}/\mathbb{Z})$ corresponding to the cuspidal group $C(\Gamma)$.

In §§2.1 - 2.3 we develop a formalism of special values (at $s = 0$ and $s = 1$) of L-functions attached to weight 2 modular forms, f. In §2.3 we define a cocycle

$$\pi_f : GL_2^+(\mathbb{Q}) \longrightarrow \mathbb{C} \quad .$$

Proposition 2.3.3 gives an explicit formula for π_f in terms of the special values of L-functions associated to f.

In §§2.4 - 2.5 we study the restriction of π to the space of Eisenstein series.

In §2.4 we study the $\mathbb{Q}$-vector space $\mathcal{E}_2(\mathbb{Q})$ of weight 2 Eisenstein series of all levels, whose q-expansions at each cusp have rational constant terms. We begin by producing a generating set

$$\{\phi_{\underline{x}}(z) \mid \underline{x} \in (\mathbb{Q}/\mathbb{Z})^2 \setminus \{\underline{0}\}\}$$

for $\mathcal{E}_2(\mathbb{Q})$. Our approach is that of Hecke ([16], §2). It should be noted that the functions $\{2\pi i \phi_{\underline{x}}\}$ are the logarithmic derivatives of the Siegel units $\{g_{\underline{x}}\}$ which play a crucial role in the theory of Kubert and Lang [21]. In Proposition 2.4.2 we prove a distribution law satisfied by the map

$$\phi : (\mathbb{Q}/\mathbb{Z})^2 \setminus \{\underline{0}\} \longrightarrow \mathcal{E}_2(\mathbb{Q})$$

$$\underline{x} \longmapsto \phi_{\underline{x}}(z) \quad .$$

This distribution law includes all of the relations amongst the $\phi_{\underline{x}}$ (2.4.4). We use the distribution law in 2.4.6 to describe the action of $GL_2(\mathbb{A})$ on $\mathcal{E}_2(\mathbb{Q})$ and in Proposition 2.4.7 we compute the action of the Hecke operators.

In §2.5 we obtain a version of a result of Schoeneberg ([40], p. 9) which expresses periods of Eisenstein series in terms of generalized Dedekind sums. In particular we find that the "real part" ξ of the cocycle π is "rational." These results yield another (more explicit) proof of Theorem 1.8.2.

§2.1. L-functions:

Let f be a weight 2 modular form of level N. We review the definition and functional equation of the L-function associated to f.

Suppose f has the Fourier expansion

$$f(z) = \sum_{n=0}^{\infty} a_n q_N^n , \quad q_N = e^{2\pi i z/N} .$$

Let $a_0(f) = a_0$ be the constant term and write $\tilde{f}(z)$ for the function $f(z) - a_0(f)$.

LEMMA 2.1.1: With f as above, and $\epsilon > 0$,

(a) $a_n = O(n^{1+\epsilon})$ as $n \to \infty$,

(b) $f(iy) = O(y^{-2-\epsilon})$ as $y \to 0$,

(c) $\tilde{f}(iy) = O(e^{-y/N})$ as $y \to \infty$.

Proof: This is well known:

(a) Hecke ([15], Satz 6).

(b) Ogg ([33], Proposition 1 Page I-3).

(c) $\lim_{y \to \infty} e^{y/N} \cdot \tilde{f}(iy) = a_1$. □

The L-function associated to f begins life as the Dirichlet series

$$L(f, s) = N^s \cdot \sum_{n=1}^{\infty} a_n n^{-s} .$$

This series converges absolutely for $Re(s) > 2$ by (a) of the lemma. The next proposition shows that $L(f, s)$ analytically continues to a meromorphic function on the whole s-plane (Ogg [33], Theorem 1 pg. I-5).

The Mellin transform of f is the function of s defined by

$$D(f, s) = \int_0^{i\infty} \tilde{f}(z)\, y^{s-1}\, dz \quad , \quad z = x + iy \quad .$$

The lemma (b), (c) shows that the integral converges absolutely for $Re(s) > 2$ and hence $D(f,s)$ defines an analytic function of s in this region.

Let

$$\sigma = \begin{pmatrix} 0 & -1 \\ 1 & 0 \end{pmatrix} \in SL_2(\mathbb{Z}) \quad ,$$

then

PROPOSITION 2.1.2:

(a) $D(f, s) = i \cdot \Gamma(s) \cdot (2\pi)^{-s} L(f, s)$

(b) $$D(f, s) = \int_i^{i\infty} \tilde{f}(z)\, y^{s-1}\, dz - \int_i^{i\infty} \widetilde{f|\sigma}(z)\, y^{1-s}\, dz$$

$$+ \, i\left(\frac{a_0(f|\sigma)}{2 - s} - \frac{a_0(f)}{s}\right) \quad .$$

(c) $D(f, s)$ can be analytically continued to a meromorphic function of s on the whole complex plane with possible simple poles only at $s = 0$ and $s = 2$, and satisfies the functional equation

$$D(f, s) + D(f|\sigma, 2-s) = 0 \quad .$$

Proof: (a) For $\mathrm{Re}(s) > 2$,

$$D(f, s) = \int_0^{i\infty} \left(\sum_{n=1}^{\infty} a_n q_N^n \right) y^{s-1} \, dz$$

$$= i \cdot \sum_{n=1}^{\infty} a_n \int_0^{\infty} e^{-2\pi n y/N} y^s \frac{dy}{y}$$

$$= i \cdot \sum_{n=1}^{\infty} \left(\frac{N}{2\pi n}\right)^s \cdot a_n \cdot \int_0^{\infty} e^{-y} y^s \frac{dy}{y}$$

$$= i \cdot \Gamma(s) \cdot \left(\frac{N}{2\pi}\right)^s \cdot \sum_{n=1}^{\infty} a_n n^{-s} \quad .$$

(b) For $\mathrm{Re}(s) > 2$,

$$D(f, s) = \int_0^{i\infty} \tilde{f}(z) y^{s-1} \, dz = \left(\int_0^i + \int_i^{i\infty} \right) \cdot (\tilde{f}(z) y^{s-1} \, dz) \quad .$$

Now

$$\int_0^i \tilde{f}(z) y^{s-1} \, dz = \int_0^i f(z) y^{s-1} \, dz - i \frac{a_0(f)}{s}$$

$$= - \int_i^{i\infty} (f|\sigma)(z) \, \mathrm{Im}(\sigma z)^{s-1} \, dz - i \frac{a_0(f)}{s}$$

$$= - \int_i^{i\infty} (\widetilde{f|\sigma})(z) \, y^{1-s} \, dz + i \frac{a_0(f|\sigma)}{2-s} - i \frac{a_0(f)}{s} \quad .$$

(c) Both integrals in (b) are absolutely convergent for all s. □

§2.2. A Calculus of Special Values:

Let f be an arbitrary weight 2 modular form of some level. Of particular interest to us are the special values of $L(f,s)$ at $s = 0$ and $s = 1$.

We have the following simple proposition.

PROPOSITION 2.2.1:

(a) $L(f, 0) = -a_0(f)$,

(b) $L(f, 1) = -2\pi i \cdot D(f, 1)$.

Proof: By Proposition 2.1.2(a)

$$D(f, s) = i \cdot \Gamma(s) \cdot (2\pi)^{-s} L(f, s) .$$

We obtain (a) by comparing the residues at $s = 0$ of both sides. Setting $s = 1$ proves (b). □

We may extend the usual action of $GL_2^+(\mathbb{Q})$ on the space of weight 2 modular forms to an action of the group ring $\mathbb{C}[GL_2^+(\mathbb{Q})]$ by linearity. For $\alpha \in \mathbb{C}[GL_2^+(\mathbb{Q})]$ write $f|\alpha$ for the result of applying α to f.

For a weight 2 modular form f define

$$e_f : \mathbb{C}[GL_2^+(\mathbb{Q})] \longrightarrow \mathbb{C}$$

by

$$e_f(\alpha) = D(f|\alpha, 1) .$$

We record the basic properties of $a_0(f)$ and e_f in the next proposition.

PROPOSITION 2.2.2: Let $\alpha, \beta \in \mathbb{C}[GL_2^+(\mathbb{Q})]$ and f be a weight 2 modular form.

(a) If $\alpha = \begin{pmatrix} a & b \\ 0 & d \end{pmatrix} \in GL_2^+(\mathbb{Q})$ then

$$a_0(f|\alpha) = \frac{a}{d} \cdot a_0(f) \quad ,$$

and

$$\widetilde{f|\alpha} = \tilde{f}\,|\,\alpha \quad .$$

(b) $e_f(\alpha\beta) = e_{f|\alpha}(\beta)$.

(c) If $\tau \in T = \{\begin{pmatrix} * & 0 \\ 0 & * \end{pmatrix}\} \subseteq GL_2^+(\mathbb{Q})$

then

$$e_f(\alpha\tau) = e_f(\alpha) \quad .$$

(d) $e_f(\alpha(1+\sigma)) = 0$,

for $\sigma = \begin{pmatrix} 0 & -1 \\ 1 & 0 \end{pmatrix} \in GL_2^+(\mathbb{Q})$.

(e) For arbitrary $z_0 \in \mathcal{H}$,

$$e_f(1) = \int_{z_0}^{i\infty} \widetilde{f|(1-\sigma)}(z)\,dz \; - \; z_0 \cdot a_0(f|(1-\sigma))$$
$$- \int_{z_0}^{\sigma z_0} f(z)\,dz \quad .$$

<u>Proof:</u> (a) $(f|\alpha)(z) = \frac{a}{d} \cdot f(\frac{az+b}{d})$, so $a_0(f|\alpha) = \frac{a}{d} \cdot a_0(f)$.

For the second part,

$$f\rceil\alpha\,(z) = (f|\alpha)(z) - a_0(f|\alpha)$$

$$= \frac{a}{d}\cdot\tilde{f}(\alpha z)$$

$$= (\tilde{f}|\alpha)(z) \quad .$$

(b) Clear.

(c) $$D(f|\alpha\tau, s) = \int_0^{i\infty} \widetilde{f|\alpha\tau}\,(z)\, y^{s-1}\, dz \;;\; \tau = \begin{pmatrix} t & 0 \\ 0 & 1 \end{pmatrix}$$

$$= \int_0^{i\infty} f\rceil\alpha\,(z)\cdot \mathrm{Im}(\tau^{-1}z)^{s-1}\, dz \quad \text{(by (a))}$$

$$= t^{1-s}\int_0^{i\infty} f\rceil\alpha\,(z)\, y^{s-1}\, dz$$

$$= t^{1-s}\, D(f|\alpha, s) \quad .$$

Now let $s = 1$.

(d) Set $s = 1$ in the functional equation

$$D(f|\alpha, s) + D(f|\alpha\sigma, 2 - s) = 0 \quad .$$

(e) The right-hand side is independent of z_0 because its derivative is 0. The result holds for $z_0 = i$ by Proposition 2.1.2(b) with $s = 1$.

□

§2.3. The Cocycle π_f and Periods of Modular Forms:

Let $\mathfrak{M}_2$ be the space of weight 2 modular forms of all levels. If we fix a $z_0 \in \mathcal{H}$ then for $\alpha \in GL_2^+(\mathbb{Q})$ we can define a linear functional

$$\varphi(\alpha) : \mathfrak{M}_2 \longrightarrow \mathbb{C}$$

by

$$\varphi_f(\alpha) = \int_{z_0}^{\alpha z_0} f(z)\, dz \quad .$$

Then φ satisfies the 1-cocycle relation

$$\varphi_f(\alpha\beta) = \varphi_{f|\alpha}(\beta) + \varphi_f(\alpha)$$

and therefore represents a cohomology class in

$$H^1(GL_2^+(\mathbb{Q}) ; \mathrm{Hom}_{\mathbb{C}}(\mathfrak{M}_2, \mathbb{C})) \quad .$$

In this section we describe another cocycle, π, which represents the same cohomology class as φ. The proposition of this section expresses π_f in terms of special values of L-functions attached to f.

Definition 2.3.1: For $f \in \mathfrak{M}_2$ define

$$\pi_f : GL_2^+(\mathbb{Q}) \longrightarrow \mathbb{C}$$

by

$$\pi_f(\alpha) = \int_{z_0}^{\alpha z_0} f(z)\, dz - z_0 \cdot a_0(f|(\alpha - 1)) + \int_{z_0}^{i\infty} \widetilde{f|(\alpha - 1)}(z)\, dz$$

for $\alpha \in GL_2^+(\mathbb{Q})$, $z_0 \in \mathcal{H}$. □

Since the derivative of the right-hand side with respect to z_0 is 0 , this definition is independent of z_0 .

Remark 2.3.2: If f is modular for a congruence group Γ and $\gamma \in \Gamma$, then

$$\pi_f(\gamma) = \int_{z_0}^{\gamma z_0} f(z)\,dz$$

is just a period of the differential form $f(z)\,dz$ on the modular curve $Y(\Gamma)$.

PROPOSITION 2.3.3:

(a) (The Cocycle Relation). For $\alpha_1, \alpha_2 \in GL_2^+(\mathbb{Q})$

$$\pi_f(\alpha_1\alpha_2) = \pi_{f|\alpha_1}(\alpha_2) + \pi_f(\alpha_1) \quad .$$

(b) If $\alpha = \begin{pmatrix} a & b \\ c & d \end{pmatrix} \in GL_2^+(\mathbb{Q})$ with $c \geq 0$, then

$$\pi_f(\alpha) = \begin{cases} \frac{a}{c}\, a_0(f) + \frac{d}{c}\, a_0(f|\alpha) - e_f\begin{pmatrix} 1 & a \\ 0 & c \end{pmatrix} & \text{if } c > 0 \ , \\ \\ \frac{b}{d}\, a_0(f) & \text{if } c = 0 \ . \end{cases}$$

(c) $\pi_f(\sigma) = -D(f, 1)$, $\sigma = \begin{pmatrix} 0 & -1 \\ 1 & 0 \end{pmatrix}$.

(d) $\pi_f\begin{pmatrix} 1 & 1 \\ 0 & 1 \end{pmatrix} = a_0(f)$. □

Proof: (c) This is Proposition 2.2.2(e).

(a) The cocycle relation is satisfied by each of the terms appearing in the definition of π_f and hence by π_f.

(b) Let $B = \{\begin{pmatrix} * & * \\ 0 & * \end{pmatrix} \in GL_2^+(\mathbb{Q})\}$ be the standard Borel subgroup. The two cases $c > 0$ and $c = 0$ correspond to the two parts of the Bruhat decomposition

$$GL_2^+(\mathbb{Q}) = B\sigma B \dot{\cup} B \ , \quad \sigma = \begin{pmatrix} 0 & -1 \\ 1 & 0 \end{pmatrix} \ .$$

First suppose $c = 0$, i.c. $\alpha \in B$. Then

$$\pi_f(\alpha) = \left(\int_{z_0}^{\alpha z_0} \tilde{f}(z)\, dz + a_0(f) \cdot (\alpha z_0 - z_0) \right)$$
$$- z_0 \cdot a_0(f|(\alpha - 1)) + \int_{z_0}^{i\infty} \widetilde{f|(\alpha - 1)}(z)\, dz$$

If z_0 tends to $i\infty$ then so does αz_0 and therefore the two integrals will tend to 0. Hence

$$\pi_f\begin{pmatrix} a & b \\ 0 & d \end{pmatrix} = \lim_{z_0 \to i\infty} \left\{ a_0(f) \cdot \left(\frac{az_0 + b}{d} - z_0 \right) - z_0 \cdot a_0(f|\alpha) + z_0 \cdot a_0(f) \right\} .$$

By Proposition 2.2.2(a) $a_0(f|\alpha) = \frac{a}{d} \cdot a_0(f)$. Simplifying

$$\pi_f(\alpha) = \lim_{z_0 \to i\infty} \left\{ a_0(f) \cdot \left(\frac{az_0 + b}{d} - z_0 - z_0 \cdot \frac{a}{d} + z_0 \right) \right\}$$
$$= \frac{b}{d} \cdot a_0(f) \quad .$$

For the case $c > 0$ we use the cocycle relation and the decomposition

$$\alpha = \begin{pmatrix} 1/c & 0 \\ 0 & 1/c \end{pmatrix} \cdot \begin{pmatrix} \delta & a \\ 0 & c \end{pmatrix} \begin{pmatrix} 0 & -1 \\ 1 & 0 \end{pmatrix} \begin{pmatrix} c & d \\ 0 & 1 \end{pmatrix} \in B\sigma B$$

with $\delta = \det(\alpha)$.

$$\pi_f(\alpha) = \pi_f\begin{pmatrix} \delta & a \\ 0 & c \end{pmatrix} + \pi_{f|\begin{pmatrix} \delta & a \\ 0 & c \end{pmatrix}}\begin{pmatrix} 0 & -1 \\ 1 & 0 \end{pmatrix} + \pi_{f|\begin{pmatrix} \delta & a \\ 0 & c \end{pmatrix}\begin{pmatrix} 0 & -1 \\ 1 & 0 \end{pmatrix}}\begin{pmatrix} c & d \\ 0 & 1 \end{pmatrix} .$$

The middle term can be calculated using (c), and the other two terms are known by the parabolic case $(c = 0)$ considered above:

$$\pi_f\begin{pmatrix} \delta & a \\ 0 & c \end{pmatrix} = \frac{a}{c} \cdot a_0(f) .$$

$$\begin{aligned} \pi_{f|\begin{pmatrix} \delta & a \\ 0 & c \end{pmatrix}\begin{pmatrix} 0 & -1 \\ 1 & 0 \end{pmatrix}}\begin{pmatrix} c & d \\ 0 & 1 \end{pmatrix} &= d \cdot a_0\left(f\Big|\begin{pmatrix} a & -\delta \\ c & 0 \end{pmatrix}\right) \\ &= d \cdot a_0\left(f\Big|\alpha \cdot \begin{pmatrix} 1 & -d \\ 0 & c \end{pmatrix}\right) \\ &= \frac{d}{c} \cdot a_0(f|\alpha) \qquad (2.2.2(a)) . \end{aligned}$$

$$\begin{aligned} \pi_{f|\begin{pmatrix} \delta & a \\ 0 & c \end{pmatrix}}\begin{pmatrix} 0 & -1 \\ 1 & 0 \end{pmatrix} &= -D\left(f\Big|\begin{pmatrix} \delta & a \\ 0 & c \end{pmatrix}, 1\right) \\ &= -D\left(f\Big|\begin{pmatrix} 1 & a \\ 0 & c \end{pmatrix}, 1\right) \qquad (2.2.2(c)) \\ &= -e_f\begin{pmatrix} 1 & a \\ 0 & c \end{pmatrix} . \end{aligned}$$

(d) Immediate from (b). □

§2.4. Eisenstein Series:

Let $\mathcal{E}_2$ denote the $\mathbb{C}$-vector space of weight 2 Eisenstein series of all levels. For a field $K \subseteq \mathbb{C}$ let $\mathcal{E}_2(K)$ be the set of $E \in \mathcal{E}_2$ such that the constant term of the q-expansion of E at each cusp is in K. Then $\mathcal{E}_2(K)$ is a K-vector space. By the discussion of §1.8 $\mathcal{E}_2(K) \cong \mathcal{E}_2(\mathbb{Q}) \otimes_{\mathbb{Q}} K$.

Let $\underline{a} = (a_1, a_2) \in (\mathbb{Q}/\mathbb{Z})^2$ and let

$$G_{\underline{a}}(z,s) = \sum_{\substack{\underline{m} \equiv \underline{a} \pmod 1 \\ \underline{m} \in \mathbb{Q}^2 \\ \underline{m} \neq \underline{0}}} (m_1 z + m_2)^{-2} \, |m_1 z + m_2|^{-s} ,$$

for $z \in \mathcal{H}$, $s \in \mathbb{C}$ with $\mathrm{Re}(s) > 0$. For fixed z, $G_{\underline{a}}(z,s)$ may be analytically continued to a meromorphic function in the s-plane which is holomorphic at $s = 0$ ([16], §2). Define

$$G_{\underline{a}}(z) = G_{\underline{a}}(z,0) \quad .$$

Then $G_{\underline{a}}(z)$ behaves like a weight 2 modular form under modular transformations of level N, but is unfortunately not quite holomorphic in z. We have (loc. cit.),

PROPOSITION 2.4.1:

a) For $\gamma \in SL_2(\mathbb{Z})$, $\underline{a} = (a_1, a_2) \in (\mathbb{Q}/\mathbb{Z})^2$

$$G_{\underline{a}} | \gamma = G_{\underline{a}\gamma}$$

where $\underline{a}\gamma = (a_1, a_2)\begin{pmatrix} a & b \\ c & d \end{pmatrix} = (aa_1 + ca_2\,,\ ba_1 + da_2)$.

b) The function

$$z \longmapsto G_{\underline{a}}(z) + \frac{2\pi i}{z - \bar{z}}$$

is holomorphic in $z \in \mathcal{H}$.

c) If $\wp_{\underline{a}}(z)$ is the $\underline{a}$-division value of the Weierstrass $\wp$-function, then

$$\wp_{\underline{a}}(z) = G_{\underline{a}}(z) - G_{\underline{0}}(z)\ ,$$

for $\underline{a} \in (\mathbb{Q}/\mathbb{Z})^2 \setminus \{\underline{0}\}$.

d) The functions

$$\{\wp_{\underline{a}}(z)\ |\ \underline{a} \in (\tfrac{1}{N}\mathbb{Z}/\mathbb{Z})^2 \setminus \{\underline{0}\}\}$$

generate the space of weight 2 Eisenstein series of level N. □

In order to obtain a basis for $\mathcal{E}_2(\mathbb{Q})$ we take for our basic Eisenstein series certain "twists" of the functions $G_{\underline{a}}(z)$.

For $\underline{x} = (x_1, x_2) \in (\frac{1}{N}\mathbb{Z}/\mathbb{Z})^2$ define the additive character

$$\psi_{\underline{x}} : (\tfrac{1}{N}\mathbb{Z}/\mathbb{Z})^2 \longrightarrow \mathbb{C}^*$$

by

$$\psi_{\underline{x}}\left(\left(\frac{a_1}{N}, \frac{a_2}{N}\right)\right) = e^{2\pi i(a_2 x_1 - a_1 x_2)}$$

Following Hecke (loc. cit.) define, for $\underline{x} \in (\frac{1}{N}\mathbb{Z}/\mathbb{Z})^2$,

$$\phi_{\underline{x}}(z) = (2\pi N)^{-2} \cdot \sum_{\underline{a} \in (\frac{1}{N}\mathbb{Z}/\mathbb{Z})^2} \psi_{\underline{x}}(\underline{a}) \cdot G_{\underline{a}}(z) \quad .$$

This definition does not depend on N. If $\underline{x} \in (\frac{1}{N}\mathbb{Z}/\mathbb{Z})^2 \setminus \{\underline{0}\}$, then $\phi_{\underline{x}}$ is a weight 2 Eisenstein series of level N. The special function $\phi_{\underline{0}}$ is not holomorphic, but is invariant for the weight 2 action of the full modular group.

Let $M_2^+(\mathbb{Z})$ be the semigroup of elements of $GL_2^+(\mathbb{Q})$ having integral entries.

PROPOSITION 2.4.2:

a) (Fourier Expansion): For $\underline{x} = (x_1, x_2) \in (\mathbb{Q}/\mathbb{Z})^2$,

$$\phi_{\underline{x}} + \delta(\underline{x}) \cdot \frac{1}{2\pi(z - \bar{z})} =$$

$$\frac{1}{2} B_2(x_1) - \sum_{\substack{k \equiv x_1 (1) \\ k \in \mathbb{Q}^+}} k \cdot \sum_{m=1}^{\infty} q(m(kz + x_2)) - \sum_{\substack{k \equiv -x_1 (1) \\ k \in \mathbb{Q}^+}} k \cdot \sum_{m=1}^{\infty} q(m(kz - x_2)) ,$$

where

$$\delta(\underline{x}) = \begin{cases} 0 & \text{if } \underline{x} \neq \underline{0} , \\ 1 & \text{if } \underline{x} = \underline{0} ; \end{cases}$$

and

$$q(w) = e^{2\pi i w} \quad \text{for} \quad w \in \mathcal{H} \quad .$$

b) (Distribution Law): Let $\alpha \in M_2^+(\mathbb{Z})$ and let $\underline{x} \in (\mathbb{Q}/\mathbb{Z})^2$, then

$$\phi_{\underline{x}} = \sum_{\substack{\underline{y} \in (\mathbb{Q}/\mathbb{Z})^2 \\ \underline{y}\alpha = \underline{x}}} \phi_{\underline{y}} \Big| \alpha \quad .$$

Sketch of Proof: (a) The Fourier expansion may be obtained with the classical methods of Poisson summation and contour integration (see for example Hecke ([16], Werke p. 469), ([14], Werke pp. 411-412), or Schoeneberg ([38], Chap. 7)). For $\underline{x} \neq \underline{0}$, the q-expansion may also be derived from the q-expansion for $\wp_{\underline{a}}(z)$.

(b) Any element, $\alpha \in GL_2^+(\mathbb{Q})$ may be expressed as a product $\alpha = \gamma\tau\gamma'$ with $\gamma, \gamma' \in SL_2(\mathbb{Z})$ and $\tau \in T = \{\begin{pmatrix} * & 0 \\ 0 & * \end{pmatrix}\}$ the standard maximal torus. If α has integral entries then so does τ. Hence we may consider the two cases $\alpha \in SL_2(\mathbb{Z})$ and $\alpha \in T$ separately.

If $\alpha \in SL_2(\mathbb{Z})$ the result follows immediately from Proposition 2.4.1(a). For $\alpha \in T$ we need to verify the following two formulae for $n \in \mathbb{Z}^+$:

(i) $$n \cdot \sum_{\substack{nx_1' \equiv x_1(1) \\ x_1' \in (\mathbb{Q}/\mathbb{Z})}} \phi_{(x_1', x_2)}(nz) = \phi_{(x_1, x_2)}(z) \quad ,$$

(ii) $$n^{-1} \cdot \sum_{\substack{nx_2' \equiv x_2(1) \\ x_2' \in (\mathbb{Q}/\mathbb{Z})}} \phi_{(x_1, x_2')}(n^{-1}z) = \phi_{(x_1, x_2)}(z) \quad .$$

Both of these are easy consequences of the Fourier expansions given in (a).

The details of (i) are as follows.

Write

$$P_{\underline{x}}(z) = \sum_{\substack{k \equiv x_1(1) \\ k \in \mathbb{Q}^+}} k \cdot \sum_{m=1}^{\infty} q(m(kz + x_2))$$

so that

$$\phi_{\underline{x}}(z) + \delta(\underline{x}) \cdot \frac{i}{2\pi(z - \bar{z})} = \frac{1}{2} B_2(x_1) - P_{\underline{x}}(z) - P_{-\underline{x}}(z) \quad .$$

The distribution law (i) holds for $P_{\underline{x}}(z)$:

$$n \cdot \sum_{\nu=0}^{n-1} P_{\left(\frac{x_1+\nu}{n}, x_2\right)}(nz) = \sum_{\nu=0}^{n-1} \sum_{\substack{k \equiv \frac{x_1+\nu}{n}(1) \\ k \in \mathbb{Q}^+}} (nk) \cdot \sum_{m=1}^{\infty} q(m(nkz + x_2))$$

$$= \sum_{\substack{k \equiv x_1(1) \\ k \in \mathbb{Q}^+}} k \cdot \sum_{m=1}^{\infty} q(m(kz + x_2))$$

$$= P_{\underline{x}}(z) \quad .$$

Since (i) also holds for the Bernoulli function $B_2(x_1)$ and for the function $\delta(\underline{x}) \cdot i/(2\pi(z - \bar{z}))$, the result follows. □

Let $S = (\mathbb{Q}/\mathbb{Z})^2$ and $S' = S \setminus \{\underline{0}\}$. Let $\mathbb{Q}[S]$ (resp $\mathbb{Q}[S']$) be the $\mathbb{Q}$-vector space generated by S (resp. S'). Let

$$\mathcal{E}_2^*(\mathbb{Q}) = \mathcal{E}_2(\mathbb{Q}) \oplus \mathbb{Q} \cdot \phi_{\underline{0}} \quad .$$

Then ϕ defines a surjective $\mathbb{Q}$-linear map

$$\phi : \mathbb{Q}[S] \longrightarrow \mathcal{E}_2^*(\mathbb{Q})$$

$$\underline{x} \in S \longmapsto \phi_{\underline{x}} ,$$

which sends $\mathbb{Q}[S']$ onto $\mathcal{E}_2(\mathbb{Q})$.

Remark 2.4.3: The Siegel units, $g_{\underline{x}}(z)$, $\underline{x} \in S'$, used by Kubert and Lang [21] in their study of the cuspidal group, may be defined by the product expansion:

$$g_{\underline{x}}(z) = q(\tfrac{z}{2} B_2(x_1)) \cdot \prod_{\substack{k \equiv x_1(1) \\ k \in \mathbb{Q}^+}} (1 - q(kz + x_2)) \cdot \prod_{\substack{k \equiv -x_1(1) \\ k \in \mathbb{Q}^+}} (1 - q(kz - x_2)) .$$

One verifies immediately, that for $\underline{x} \in S'$

$$2\pi i \cdot \phi_{\underline{x}}(z) = g'_{\underline{x}}(z) / g_{\underline{x}}(z) .$$

Furthermore, for the special case $\underline{x} = \underline{0}$, we have

$$2\pi i \cdot \phi_{\underline{0}}(z) - \frac{1}{z - \bar{z}} = 2 \cdot \eta'(z)/\eta(z) ,$$

where $\eta(z)$ is the Dedekind η-function. $\sqcap$

Remark 2.4.4: The map $\phi : (\mathbb{Q}/\mathbb{Z})^2 \to \mathcal{E}_2^*(\mathbb{Q})$ is the universal even distribution on $(\mathbb{Q}/\mathbb{Z})^2$ in the sense of Kubert ([20], especially Cor. 4.15).

To see this, define first an action of $M_2^+(\mathbb{Z})$ on $\mathbb{Q}[S]$ by: for $\underline{x} \in S$, $\alpha \in M_2^+(\mathbb{Z})$,

$$\underline{x}|\alpha \overset{\text{dfn}}{=\!=} \sum_{\substack{\underline{y}\alpha' = \underline{x} \\ y \in S}} \underline{y} \in \mathbb{Q}[S] ,$$

where $\alpha' = \det(\alpha) \cdot \alpha^{-1} \in M_2^+(\mathbb{Z})$. With this action we have

$$\phi_{\underline{x}|\alpha} = \phi_{\underline{x}}|\alpha \qquad \text{for} \qquad \alpha \in M_2^+(\mathbb{Z}) \quad .$$

The subspace

$$\underline{\text{Dist}} = \mathbb{Q}[\underline{x} - \underline{x}|\begin{pmatrix} n & 0 \\ 0 & n \end{pmatrix} : \underline{x} \in S \ , \ n \in \mathbb{Z}\setminus\{0\}]$$
$$\subseteq \mathbb{Q}[S]$$

is the space of even distribution relations used by Kubert (loc. cit). Clearly $\underline{\text{Dist}} \subseteq \text{Ker}(\phi)$, so ϕ is an even distribution. That ϕ is the universal even distribution is now just a dimension count. The image under ϕ of $(\frac{1}{N}\mathbb{Z}/\mathbb{Z})^2$ is the number of cusps on $X(N)$. But Kubert (loc. cit) has shown that this is also the dimension of the universal even distribution on $(\frac{1}{N}\mathbb{Z}/\mathbb{Z})^2$.

□

Remark 2.4.5: Since $PGL_2^+(\mathbb{Q}) \cong M_2^+(\mathbb{Z})/\{\text{scalars}\}$, we have an action of $PGL_2^+(\mathbb{Q})$ on $\mathbb{Q}[S]/\underline{\text{Dist}}$ for which ϕ is $PGL_2^+(\mathbb{Q})$-equivariant. Hence, the map

$$\phi : \mathbb{Q}[S]/\underline{\text{Dist}} \longrightarrow \mathcal{E}_2^*(\mathbb{Q})$$

is a $GL_2^+(\mathbb{Q})$-isomorphism. □

Remark 2.4.6: The distribution law can be used to define the action of $GL_2(\mathbb{A})$ on $\mathcal{E}_2^*(\mathbb{Q})$ where $\mathbb{A}$ is the group of adéles for $\mathbb{Q}$. Letting $GL_2(\mathbb{R})$ act trivially and $\prod_p GL_2(\mathbb{Z}_p)$ act componentwise on

$$(\mathbb{Q}/\mathbb{Z})^2 \cong \coprod_p (\mathbb{Q}_p/\mathbb{Z}_p)^2 \quad ,$$

yields an action of the subgroup

$$U = GL_2(\mathbb{R}) \times \prod_p GL_2(\mathbb{Z}_p) \subseteq GL_2(\mathbb{A})$$

on $\mathbb{Q}[S]$. This action preserves Dist = Ker(ϕ) so we have a homomorphism

$$\tau : U \longrightarrow \mathrm{Aut}_{\mathbb{Q}}(\mathcal{E}_2^*(\mathbb{Q})) \quad .$$

We have $GL_2(\mathbb{A}) = U \cdot GL_2^+(\mathbb{Q})$. Furthermore, if $\alpha, \beta \in GL_2^+(\mathbb{Q})$, and $u, v \in U$ satisfy $\alpha u = v\beta$, then an easy calculation with the distribution law 2.4.2(b) shows

$$(\phi_{\underline{x}} | \alpha) \mid \tau(u) = (\phi_{\underline{x}} | \tau(v)) \mid \beta, \qquad \underline{x} \in S \quad .$$

Hence, τ extends to a homomorphism

$$\tau : GL_2(\mathbb{A}) \longrightarrow \mathrm{Aut}_{\mathbb{Q}}(\mathcal{E}_2^*(\mathbb{Q}))$$

which extends the usual action of $GL_2^+(\mathbb{Q})$ on $\mathcal{E}_2^*(\mathbb{Q})$. □

The distribution law may also be used to compute the action of the Hecke operators on $\mathcal{E}_2(\mathbb{Q})$. Fix N_1 and N_2 to be positive integers and let $N = \mathrm{lcm}(N_1, N_2)$. Let $a, b \in \mathbb{Z}$ and set $\underline{x} = (a/N_2, b/N_1) \in (\mathbb{Q}/\mathbb{Z})^2$. Then $\phi_{\underline{x}}$ is modular for $\Gamma = \Gamma_1(N_1, N_2)$.

For a prime ℓ define the operators $U_\ell^{(N_2)}$, π_ℓ by

$$U_\ell^{(N_2)} = \sum_{k=0}^{\ell-1} \begin{pmatrix} 1 & N_2 k \\ 0 & \ell \end{pmatrix} ,$$

$$\pi_\ell = \begin{pmatrix} \ell & 0 \\ 0 & 1 \end{pmatrix} \quad .$$

If $\ell \nmid N$ let $\sigma_\ell^{(N)} \in SL_2(\mathbb{Z})$ be such that

$$\sigma_\ell^{(N)} \equiv \begin{pmatrix} * & 0 \\ 0 & \ell \end{pmatrix} \pmod N \quad .$$

For $\underline{x}$ and Γ as above the action of the Hecke operators T_ℓ on $\phi_{\underline{x}}$ is given by

$$\phi_{\underline{x}}\big|T_\ell = \begin{cases} \phi_{\underline{x}}\big|U_\ell^{(N_2)} + \phi_{\underline{x}}\big|\sigma_\ell^{(N)} \cdot \pi_\ell , & \text{if } \ell \nmid N \ ; \\ \phi_{\underline{x}}\big|U_\ell^{(N_2)} & \text{if } \ell \mid N \ ; \end{cases}$$

(see §1.2).

PROPOSITION 2.4.7: (Hecke operators): Let $\underline{x} = (x, y) = (a/N_2, b/N_1) \in (\mathbb{Q}/\mathbb{Z})^2$ with $(a, N_2) = 1$. Let ℓ be a prime. If $\ell \nmid N_2$, let $\ell' \in \mathbb{Z}$ be such that $\ell\ell' \equiv 1 \pmod{N_2}$. Then

$$\phi_{\underline{x}}\big|T_\ell = \begin{cases} \ell \cdot \phi_{(\ell' x, y)} + \phi_{(x, \ell y)} & \text{if } \ell \nmid N , \\ \phi_{(x, \ell y)} & \text{if } \ell \mid N_2 \\ \phi_{(x, \ell y)} + \ell \cdot \phi_{(\ell' x, y)} - \sum_{k=0}^{\ell - 1} \phi_{(\ell' x, y + \frac{k}{\ell})} & \text{if } \ell \nmid N_2 \text{ but } \ell \mid N . \end{cases}$$

□

Proof: We compute the action of $U_\ell^{(N_2)}$ and $\sigma_\ell^{(N)} \cdot \pi_\ell$ on $\phi_{\underline{x}}$ separately.

$$\phi_{\underline{x}}\Big|U_\ell^{(N_2)} = \sum_{k=0}^{\ell-1} \phi_{\underline{x}}\Big|\begin{pmatrix}1 & 0\\ 0 & \ell\end{pmatrix}\begin{pmatrix}1 & N_2 k\\ 0 & 1\end{pmatrix}$$

$$= \sum_{k=0}^{\ell-1}\sum_{r=0}^{\ell-1} \phi_{\left(\frac{\frac{a}{N_2}+r}{\ell},\ \frac{b}{N_1}\right)}\Bigg|\begin{pmatrix}1 & N_2 k\\ 0 & 1\end{pmatrix} \qquad \text{(distribution law)}$$

$$= \sum_{k=0}^{\ell-1}\sum_{r=0}^{\ell-1} \phi_{\left(\frac{a+N_2 r}{N_2 \ell},\ \frac{b}{N_1}+\frac{k(a+N_2 r)}{\ell}\right)} .$$

To simply this last expression we will use the distribution law and the relation $\phi_{\underline{x}}\Big|\begin{pmatrix}\ell & 0\\ 0 & \ell\end{pmatrix} = \phi_{\underline{x}}$.

If $\ell \mid N_2$ we may do this directly. In this case $a + N_2 r$ is prime to ℓ for each r and hence the above sum is just

$$\phi_{\underline{x}}\Big|U_\ell^{(N_2)} = \sum_{k=0}^{\ell-1}\sum_{r=0}^{\ell-1} \phi_{\left(\frac{x+r}{\ell},\ y+\frac{k}{\ell}\right)}$$

$$= \phi_{(x,\ell y)} \qquad \text{(distribution law)} .$$

This is the second case of the proposition.

If $\ell \nmid N_2$ we can still use the distribution law but will have terms left over:

$$\phi_{\underline{x}}\Big|U_{\ell}^{(N_2)} = \sum_{\substack{r=0 \\ a+N_2 r \not\equiv 0(\ell)}}^{\ell-1} \sum_{k=0}^{\ell-1} \phi_{\left(\frac{x+r}{\ell},\, y+\frac{k}{\ell}\right)} + \ell \cdot \phi_{(\ell' x,\, y)}$$

$$= \sum_{r=0}^{\ell-1} \sum_{k=0}^{\ell-1} \phi_{\left(\frac{x+r}{\ell},\, y+\frac{k}{\ell}\right)} - \sum_{k=0}^{\ell-1} \phi_{(\ell' x,\, y+\frac{k}{\ell})} + \ell \cdot \phi_{(\ell' x,\, y)}$$

$$= \phi_{(x,\ell y)} + \ell \cdot \phi_{(\ell' x,\, y)} - \sum_{k=0}^{\ell-1} \phi_{(\ell' x,\, y+\frac{k}{\ell})} ,$$

which gives us the third case of the proposition.

To compute $\phi_{\underline{x}}\Big|\sigma_{\ell}^{(N)} \cdot \pi_{\ell}$ when $\ell \nmid N$ we use the distribution law again:

$$\phi_{\underline{x}}\Big|\sigma_{\ell}^{(N)} \cdot \pi_{\ell} = \phi_{(\ell' x,\, \ell y)}\Big|\begin{pmatrix} \ell & 0 \\ 0 & 1 \end{pmatrix}$$

$$= \sum_{k=0}^{\ell-1} \phi_{(\ell' x,\, y+\frac{k}{\ell})} .$$

Adding this to the above formula for $\phi_{\underline{x}}\Big|U_{\ell}^{(N_2)}$ gives us the first equality of the proposition. □

§2.5. Periods of Eisenstein Series:

It is natural to separate the examples of our construction (§§2.1 - 2.3) into two classes: the cusp forms and the Eisenstein series. The case of cusp forms is still quite mysterious. The Eisenstein series on the other hand are very well understood. The possibility of congruences between the two was discussed in Chapter 1.

In this section we restrict our 1-cocycle π to the space $\mathcal{E}_2$ of Eisenstein series. We will construct a "rational" cocycle

$$\xi \in Z^1(GL_2^+(\mathbb{Q})\, ;\, \mathrm{Hom}_{\mathbb{Q}}(\mathcal{E}_2(\mathbb{Q})\, ;\, \mathbb{Q}))$$

which is the "real part" of π. Proposition 2.5.4 gives explicit formulae for

$$\xi_{\phi_{\underline{x}}} : GL_2^+(\mathbb{Q}) \longrightarrow \mathbb{Q} \quad .$$

In particular we obtain the classical formulae for the periods of Eisenstein series (compare Schoeneberg ([40], pg. 9)). The main term in these formulae involves the generalized Dedekind sum first introduced by Meyer [31], [32]. As a consequence we obtain another proof that the periods of an Eisenstein series lie in the $\mathbb{Q}$-span of the constant terms of its q-expansions (compare Theorem 1.8.2).

For $x \in \mathbb{Q}/\mathbb{Z}$ define the following functions of the complex variable s.

$$Z(s, x) = \sum_{n=1}^{\infty} e^{2\pi i n x} n^{-s} ,$$

$$\zeta(s, x) = \sum_{\substack{k \equiv x(1) \\ k \in \mathbb{Q}^+}} k^{-s} .$$

Recall the functions $\phi_{\underline{x}} \in \mathcal{E}_2^*(\mathbb{Q})$ for $\underline{x} \in (\mathbb{Q}/\mathbb{Z})^2$ from §2.4.

PROPOSITION 2.5.1:

(a) (Mellin Transform). For $\underline{x} \neq \underline{0}$,

$$D(\phi_{\underline{x}}, s) = -i\,\Gamma(s) \cdot (2\pi)^{-s} \cdot \{Z(s, x_2) \cdot \zeta(s-1, x_1) + Z(s, -x_2)\, \zeta(s-1, -x_1)\}.$$

(b) (Special Values). For $\underline{x} \neq \underline{0}$,

i) $$D(\phi_{\underline{x}}, 1) = \mathbf{B}_1(x_1) \cdot \mathbf{B}_1(x_2) + \frac{1}{2\pi i}\left\{\delta_{x_1} \cdot \left(\log\left|1 - e^{2\pi i x_2}\right|\right) - \delta_{x_2} \cdot \left(\log\left|1 - e^{2\pi i x_1}\right|\right)\right\}$$

where

$$\delta_x = \begin{cases} 1 & \text{if} \quad x \equiv 0 \pmod 1 \\ 0 & \text{if} \quad x \not\equiv 0 \pmod 1 \end{cases} .$$

ii) $$a_0(\phi_{\underline{x}}) = \frac{1}{2} \mathbf{B}_2(x_1) .$$

Proof: (a) For $\mathrm{Re}(s) > 2$ we can compute the L-function directly from the q-expansion for $\phi_{\underline{x}}$ given in 2.4.2(a). Let

$$P_{\underline{x}}(z) = \sum_{\substack{k \equiv x_1 (1) \\ k \in \mathbb{Q}^+}} k \cdot \sum_{m=1}^{\infty} q(m(kz + x_2)) \quad .$$

Then

$$\phi_{\underline{x}}(z) = \frac{1}{2} \mathbf{B}_2(x_1) - P_{\underline{x}}(z) - P_{\underline{-x}}(z) \quad ,$$

and

$$L(\phi_{\underline{x}}, s) = - L(P_{\underline{x}}, s) - L(P_{\underline{-x}}, s) \quad .$$

We have

$$L(P_{\underline{x}}, s) = \sum_{\substack{k \equiv x_1 (1) \\ k \in \mathbb{Q}^+}} k \cdot \sum_{m=1}^{\infty} q(m x_2) (m k)^{-s}$$

$$= \left(\sum_{\substack{k \equiv x_1 (1) \\ k \in \mathbb{Q}^+}} k^{1-s} \right) \cdot \left(\sum_{m=1}^{\infty} e^{2\pi i m x_2} m^{-s} \right)$$

$$= \zeta(s-1, x_1) \cdot Z(s, x_2) \quad .$$

Since $D(\phi_{\underline{x}}, s) = i \cdot \Gamma(s) (2\pi)^{-s} L(\varphi_{\underline{x}}, s)$ the result follows.

(b) The value of the constant term of $\varphi_{\underline{x}}$ is given in 2.4.2(a).

To calculate $D(\phi_{\underline{x}}, 1)$ we use the following formulae: for $x \in \mathbb{Q}/\mathbb{Z}$,

(i) $\zeta(0, x) = - \mathbf{B}_1(x) - \frac{1}{2} \delta_x$.

([47], pg. 271).

(ii) $Z(1,x) - Z(1,-x) = -2\pi i\, B_1(x)$ if $x \neq 0$, ([36], pg. 16).

(iii) $Z(1,x) + Z(1,-x) = -2 \cdot \log |1 - e^{2\pi i x}|$ if $x \neq 0$.

If $x_1, x_2 \neq 0$ then we have

$$D(\phi_{\underline{x}}, 1) = \frac{1}{2\pi i} \{Z(1,x_2)\zeta(0,x_1) + Z(1,-x_2)\zeta(0,-x_1)\}$$

$$= \frac{1}{2\pi i} \{2\pi i \cdot B_1(x_1) \cdot B_1(x_2)\} \quad .$$

If $x_1 = 0$, $x_2 \neq 0$ then $B_1(x_1) = 0$ and

$$D(\phi_{\underline{x}}, 1) = \frac{1}{2\pi i} \cdot \{\frac{-1}{2} (Z(1,x_2) + Z(1,-x_2))\} \quad .$$

For the case $x_1 \neq 0$, $x_2 = 0$ use the functional equation

$$D(\phi_{\underline{x}}, 1) + D(\phi_{\underline{x}\sigma}, 1) = 0, \qquad \sigma = \begin{pmatrix} 0 & -1 \\ 1 & 0 \end{pmatrix}$$

and apply the last case. □

With this proposition and 2.3.3(b) we can now give a formula for the cocycle π restricted to the set $\{\phi_{\underline{x}} \mid \underline{x} \in S'\}$.

A look at the form of $D(\phi_{\underline{x}}, 1)$ in the last proposition suggests that we should decompose π into its "real" and "imaginary" parts. Let

$$\iota = \begin{pmatrix} -1 & 0 \\ 0 & 1 \end{pmatrix} \in GL_2(\mathbb{A}) ,$$

$$\pi^{\iota} \in Z^1(GL_2^+(\mathbb{Q}) ; \mathrm{Hom}_{\mathbb{C}}(\mathcal{E}_2, \mathbb{C}))$$

by: for $E \in \mathcal{E}_2$, $\alpha \in GL_2^+(\mathbb{Q})$,

$$\pi_E^{\iota}(\alpha) = -\pi_{E|T(\iota)}(\iota^{-1}\alpha\iota) \quad .$$

Let

$$\xi = \frac{1}{2}(\pi + \pi^{\iota})$$

and

$$\xi' = \frac{1}{2i}(\pi - \pi^{\iota}) \quad .$$

Then

$$\pi = \xi + i \cdot \xi' \quad .$$

For $E \in \mathcal{E}_2$ define $e_E^{\iota} : GL_2^+(\mathbb{Q}) \to \mathbb{C}$ by

$$e_E^{\iota}(\alpha) = -e_{E|T(\iota)}(\iota^{-1}\alpha\iota) \quad .$$

<u>Definition 2.5.2</u>: The <u>Dedekind symbol</u> associated to an Eisenstein series $E \in \mathcal{E}_2$ is the symbol

$$s_E : \mathbb{P}^1(\mathbb{Q}) \longrightarrow \mathbb{C}$$

defined by

$$s_E(r) = \begin{cases} \frac{1}{2}\left(e_E\begin{pmatrix}1 & r\\ 0 & 1\end{pmatrix} + e_E^{\iota}\begin{pmatrix}1 & r\\ 0 & 1\end{pmatrix}\right) & \text{if } r \in \mathbb{Q} \, , \\ 0 & \text{if } r = i\infty \, . \end{cases}$$

□

Define $s'_E : \mathbb{P}^1(\mathbb{Q}) \to \mathbb{C}$ by setting $s'_E(i\infty) = 0$ and letting $s'_E(r)$ satisfy the equation

$$e_E\begin{pmatrix}1 & r\\ 0 & 1\end{pmatrix} = s_E(r) + i \cdot s'_E(r)$$

for $r \in \mathbb{Q}$.

By 2.3.3(b) we have for $\alpha = \begin{pmatrix}a & b\\ c & d\end{pmatrix} \in GL_2^+(\mathbb{Q})$ $c \geq 0$.

$$\xi_E(\alpha) = \begin{cases} \frac{a}{c} \cdot a_0(E) + \frac{d}{c}\, a_0(E|\alpha) - s_E(\frac{a}{c}) & \text{if } c > 0 , \\ \frac{b}{d} \cdot a_0(E) & \text{if } c = 0 ; \end{cases}$$

(2.5.3)

and

$$\xi'_E(\alpha) = - s'_E(\tfrac{a}{c}) \quad .$$

We give explicit formulae for the symbols $s_{\phi_{\underline{x}}}$, $s'_{\phi_{\underline{x}}}$, $\underline{x} \in S'$ in the next proposition. Since the functions $\phi_{\underline{x}}$, $\underline{x} \in S'$ span $\mathcal{E}_2$, this gives us an explicit formula for the cocycle π.

PROPOSITION 2.5.4: Let $\underline{x} = (x_1, x_2) \in S'$, $m, n \in \mathbb{Z}$ with $(m, n) = 1$ and $n > 0$.

(a)

$$s_{\phi_{\underline{x}}}(\tfrac{m}{n}) = \sum_{\nu=0}^{n-1} \mathbf{B}_1\left(\frac{x_1 + \nu}{n}\right) \cdot \mathbf{B}_1\left(m\,\frac{x_1+\nu}{n} + x_2\right)$$

(so $s_{\phi_{\underline{x}}}$ coincides with Meyer's generalized Dedekind sum $s_{\underline{x}}$ ([31]

pg. 174), [32]).

(b) Define $F : S' \to \mathbb{C}$ by

$$F(\underline{x}) = -\delta_{x_1} \cdot \log \left| 1 - e^{2\pi i x_2} \right| .$$

Then

$$s'_{\phi_{\underline{x}}}\left(\frac{m}{n}\right) = \frac{1}{2\pi}\left(F(\underline{x}\,\gamma) - F(\underline{x})\right) ,$$

where $\gamma \in SL_2(\mathbb{Z})$ is chosen so that $\gamma \cdot (i\infty) = \left(\frac{m}{n}\right)$.

Proof: (a) By 2.5.1(b) we find

$$s_{\phi_{\underline{x}}}\left(\frac{0}{1}\right) = \mathbf{B}_1(x_1) \cdot \mathbf{B}_1(x_2) .$$

Also
$$s_{\phi_{\underline{x}}}\left(\frac{m}{n}\right) = s_{\phi_{\underline{x}}|\begin{pmatrix}1 & m\\ 0 & n\end{pmatrix}}\left(\frac{0}{1}\right) .$$

An application of the distribution law 2.4.2(b) gives

$$\begin{aligned}
\phi_{\underline{x}}\left|\begin{pmatrix}1 & m\\ 0 & n\end{pmatrix}\right. &= \phi_{\underline{x}}\left|\begin{pmatrix}1 & 0\\ 0 & n\end{pmatrix}\begin{pmatrix}1 & m\\ 0 & 1\end{pmatrix}\right. \\
&= \sum_{\nu=0}^{n-1} \phi_{\left(\frac{x_1+\nu}{n},\, x_2\right)}\left|\begin{pmatrix}1 & m\\ 0 & 1\end{pmatrix}\right. \\
&= \sum_{\nu=0}^{n-1} \phi_{\left(\frac{x_1+\nu}{n},\, m \cdot \frac{x_1+\nu}{n} + x_2\right)} .
\end{aligned}$$

(b) We need to show $\xi'_{\phi_{\underline{x}}}(\gamma) = \frac{1}{2\pi}(F(\underline{x}\gamma) - F(\underline{x}))$ for $\gamma \in SL_2(\mathbb{Z})$.

Both sides of this supposed equality are 1-cocycles when viewed as functions of $\gamma \in SL_2(\mathbb{Z})$. So we only need to check it on a set of generators for $SL_2(\mathbb{Z})$:

$$\xi'_{\phi_{\underline{x}}}\begin{pmatrix} 0 & -1 \\ 1 & 0 \end{pmatrix} = -s'_{\phi_{\underline{x}}}\left(\frac{0}{1}\right)$$

$$= \frac{-1}{2\pi}\left(\delta_{x_2} \cdot \log\left|1 - e^{2\pi i x_1}\right| - \delta_{x_1} \cdot \log\left|1 - e^{2\pi i x_2}\right|\right) \qquad (2.5.1(b))$$

$$= \frac{1}{2\pi}\left(F\left(\underline{x}\begin{pmatrix} 0 & -1 \\ 1 & 0 \end{pmatrix}\right) - F(\underline{x})\right) \; ;$$

and

$$\xi'_{\phi_{\underline{x}}}\begin{pmatrix} 1 & 1 \\ 0 & 1 \end{pmatrix} = -s'_{\phi_{\underline{x}}}(i\infty) = 0 = \frac{1}{2\pi}\left(F\left(\underline{x}\begin{pmatrix} 1 & 1 \\ 0 & 1 \end{pmatrix}\right) - F(\underline{x})\right) .$$

□

For a $\mathbb{Q}$-vector space V write V^* for $\mathrm{Hom}_{\mathbb{Q}}(V, \mathbb{C})$. By Remark 2.4.4 we have an exact sequence

$$0 \longrightarrow \mathcal{E}_2(\mathbb{Q})^* \longrightarrow \mathbb{Q}[S]^* \longrightarrow \underline{\mathrm{Dist}}^* \longrightarrow 0$$

of $SL_2(\mathbb{Z})$-modules. Part (b) of the last proposition shows that the cohomology class, $\bar{\xi}'$, represented by ξ' in $H^1(SL_2(\mathbb{Z}); \mathcal{E}_2(\mathbb{Q})^*)$ has zero image in $H^1(SL_2(\mathbb{Z}); \mathbb{Q}[S]^*)$. Nevertheless it is not difficult to show $\bar{\xi}' \neq 0$. At least we have the following.

COROLLARY 2.5.5: Let Γ be a congruence group and $E \in \mathcal{E}_2(\Gamma)$. Then

$$\xi'_E(\gamma) = 0$$

for $\gamma \in \Gamma$. Hence $\pi_E(\gamma) = \xi_E(\gamma)$.

Proof: Suppose Γ has level $N > 0$, and let $\gamma \in \Gamma(N)$. The Eisenstein series E is a linear combination of the functions $\phi_{\underline{x}}$, $x \in (\frac{1}{N}\mathbb{Z}/\mathbb{Z})^2 \backslash \{\underline{0}\}$. Since for each such $\phi_{\underline{x}}$

$$\xi_{\phi_{\underline{x}}}(\gamma) = \frac{1}{2\pi}(F(\underline{x}\gamma) - F(\underline{x})) = 0$$

we also have $\xi'_E(\gamma) = 0$.

If β is an arbitrary element of Γ then $\beta^n \in \Gamma(N)$ for some n. Then

$$\xi'_E(\beta) = \frac{1}{n}\,\xi'_E(\beta^n) = 0 \quad . \qquad \square$$

COROLLARY 2.5.6: Let K be a number field and $E \in \mathcal{E}_2(K)$. Then ξ_E is K-valued:

$$\xi_E : GL_2^+(\mathbb{Q}) \longrightarrow K \quad .$$

Proof: E is a K-linear combination of the functions $\phi_{\underline{x}}$, $\underline{x} \in S'$. But by (2.5.3) and Proposition 2.5.4(a), $\xi_{\phi_{\underline{x}}}$ is $\mathbb{Q}$-valued for $\underline{x} \in S'$. $\square$

If $E \in \mathcal{E}_2(\Gamma;\mathbb{Q})$ for a congruence group Γ, then Periods (E) = $\pi_E(\Gamma) = \xi_E(\Gamma) \subseteq \mathbb{Q}$ by the last two corollaries. This provides another proof of Theorem 1.8.2.

Chapter 3. The Special Values Associated to Cuspidal Groups

Let Γ be a group of type (N_1, N_2) and $E \in \mathcal{E}_2(\Gamma)$ be a $\mathfrak{T}$-eigenfunction. This chapter is devoted to describing the subgroup C_E of the cuspidal group and the speical values of the associated cohomology class $\varphi_E \in H^1(X; A(E))$.

The main result of §3.1 is Theorem 3.1.2 which says that if E is an Eisenstein series and $\chi \neq 1$ a primitive Dirichlet character of conductor prime to N_1N_2, then the special value $\Lambda(\varphi_E, \chi)$ is essentially $\frac{\tau(\bar{\chi})}{2\pi i} \cdot L(E, \chi, 1)$ reduced modulo the residues $R(E)[\chi]$. There are, however, two extra terms in the formula. These will be non-zero only if χ is in a certain exceptional set (Definition 3.1.3 and Corollary 3.1.5).

In §3.2 we study $\mathfrak{T}$-eigenfunctions, $E \in \mathcal{E}_2(\Gamma)$, and the Galois module structure of the associated subgroups, C_E, of the cuspidal group. Proposition 3.2.2 shows that there are Dirichlet characters ϵ_1, ϵ_2 such that the eigenvalue of T_ℓ acting on E is $(\epsilon_1(\ell) + \ell\epsilon_2(\ell))$ for each prime $\ell \nmid N_1N_2$. Theorem 3.2.4 shows that C_E is a cyclic $\mathbb{Z}[\epsilon_1, \epsilon_2]$-module on which $\mathrm{Gal}(\overline{\mathbb{Q}}/\mathbb{Q})$ acts via ϵ_1.

We collect some standard facts concerning Dirichlet L-functions in §3.3. These include the functional equation and special values at nonpositive integers.

In §3.4 we associate to a pair of Dirichlet characters $\varepsilon_i : (\mathbb{Z}/N_i\mathbb{Z})^* \to \mathbb{C}^*$, $i = 1,2$, an Eisenstein series $E = E(\varepsilon_1, \varepsilon_2) \in \mathcal{E}_2(\Gamma_1(N_1, N_2))$. This Eisenstein series is an eigenfunction for the Hecke operators T_ℓ, $\ell \nmid N_1N_2$ with Nebentypus character $\varepsilon = \varepsilon_1\varepsilon_2$. The L-function of E is $L(E, s) = -2N_2^{s-1} \cdot L(\hat{\bar{\varepsilon}}_1, s) \cdot L(\varepsilon_2, s-1)$ (Proposition 3.4.2). The special values $\Lambda(\varphi_E, \chi)$ are given in Corollary 3.4.3. Proposition 3.4.4 gives the relevant information for describing the group $R(E)$ and the cohomology class φ_E.

In §3.5 we use a result of E. Friedman to prove a theorem concerning the nonvanishing of the special values $\Lambda(\varphi_E, \chi)$ where $E = E(\varepsilon_1, \varepsilon_2)$ and $\varepsilon_1, \varepsilon_2$ are primitive.

Finally in §3.6 we prove a result which is helpful in applications for computing the group of periods of $E(\varepsilon_1, \varepsilon_2)$ for arbitrary $\varepsilon_1, \varepsilon_2$.

§3.1. Special Values Associated to the Cuspidal Group:

Let Γ be a congruence group of type (N_1, N_2), $X = X(\Gamma)$ and $C = C(\Gamma)$ the cuspidal group. Let χ be a primitive Dirichlet character of conductor $m > 1$ prime to $N = \text{lcm}(N_1, N_2)$.

The theorem of this section describes the special values $\Lambda(\varphi_E, \chi)$, $E \in \mathcal{E}_2(\Gamma)$. The main term of this formula is $L(E, \chi, 1)$ multiplied by a Gauss sum. There are also remainder terms which involve residues of $\omega(E)$ on X and special values of the Dirichlet L-function $L(\chi, s)$.

The following lemma extends Proposition 1.6.3 to include the case where f is an Eisenstein series. The integral appearing in 1.6.3 does not converge in this case.

LEMMA 3.1.1: Let f be a weight two modular form for Γ. Then

$$\tau(\bar{\chi})\, L(f, \chi, 1) = -2\pi i \sum_{a=0}^{m-1} \bar{\chi}(aN_2)\, e_f\begin{pmatrix} 1 & aN_2 \\ 0 & m \end{pmatrix} .$$

(Notation as in §2.3.)

Proof: As in §1.6 set

$$f_\chi(z) = \bar{\chi}(N_2) \cdot \sum_{n=0}^{\infty} a_n\, \chi(n)\, q_{N_2}^n .$$

Then

$$\tau(\bar{\chi}) \cdot f_\chi(z) = \sum_{a=0}^{m-1} \bar{\chi}(aN_2) \cdot f\left(z + \frac{aN_2}{m}\right) .$$

Hence

$$\tau(\overline{\chi}) \cdot D(f_\chi, s) = \sum_{a=0}^{m-1} \overline{\chi}(aN_2) \cdot D\left(f \,\middle|\, \begin{pmatrix} 1 & aN_2/m \\ 0 & 1 \end{pmatrix}, s\right) ,$$

and finally

$$\tau(\overline{\chi}) \cdot L(f, \chi, 1) = -2\pi i \cdot \tau(\overline{\chi}) \cdot D(f_\chi, 1)$$

$$= -2\pi i \cdot \sum_{a=0}^{m-1} \overline{\chi}(aN_2) \cdot e_f\begin{pmatrix} 1 & aN_2 \\ 0 & m \end{pmatrix} .$$

□

Let K be a number field, and $E \in \mathcal{E}_2(\Gamma; K)$. In §1.8 we associated to E the finite group

$$A(E) = \underline{\text{Periods}}(E)/R(E)$$

and the cohomology class

$$\varphi_E \in H^1(X; A(E)) .$$

Let $\underline{\text{Periods}}(E)[\chi]$ (resp. $R(E)[\chi]$) be the $\mathbb{Z}[\chi]$-module generated by $\underline{\text{Periods}}(E)$ (resp. $R(E)$) in $K \cdot \mathbb{Q}[\chi]$. There is a natural surjective $\mathbb{Z}[\chi]$-homomorphism

$$A(E) \otimes_{\mathbb{Z}} \mathbb{Z}[\chi] \twoheadrightarrow A(E)[\chi] \overset{\text{dfn}}{=\!=} \underline{\text{Periods}}(E)[\chi]/R(E)[\chi] .$$

The next theorem describes the special values

$$\Lambda(\varphi_E, \chi) \in A(E)[\chi] .$$

Recall (§1.2) the matrices $\sigma_m \in SL_2(\mathbb{Z})$ for $m \in (\mathbb{Z}/N\mathbb{Z})^*$.

They satisfy the congruences

$$\sigma_m \equiv \begin{pmatrix} * & 0 \\ 0 & m \end{pmatrix} \pmod{N} \quad .$$

Let $\sigma = \begin{pmatrix} 0 & -1 \\ 1 & 0 \end{pmatrix}$.

THEOREM 3.1.2: Let $E \in \mathcal{E}_2(\Gamma ; K)$. Then

$$\Lambda(\varphi_E, \chi) \equiv \bar{\chi}(N_2) \cdot B_1(\bar{\chi}) \cdot N_2 a_0(E) + \chi(N_1) \cdot B_1(\chi) \cdot N_1 a_0(E|\sigma_m \sigma)$$

$$+ \frac{\tau(\bar{\chi})}{2\pi i} \cdot L(E, \chi, 1)$$

$$(\mathrm{mod}\ R(E)[\chi]) \quad .$$

Proof: Let $\beta = \sigma_m \sigma \in SL_2(\mathbb{Z})$. For $a \in \mathbb{Z}$ with $(a, m) = 1$, the cusps $(a N_2/m)$ and $\beta \cdot (i\infty)$ are Γ-equivalent. Let $\gamma_a \in \Gamma$ such that $\gamma_a \beta \cdot (i\infty) = (a N_2/m)$. Then

$$\Lambda(\varphi_E, \chi) = \varphi_E \cap \Lambda(\chi) = \varphi_E \cap \sum_{a=0}^{m-1} \bar{\chi}(a N_2) \cdot \left\{ i\infty , \frac{a N_2}{m} \right\}_\Gamma$$

$$= \varphi_E \cap \sum_{a=0}^{m-1} \bar{\chi}(a N_2) \cdot \left\{ \beta \cdot (i\infty) , \frac{a N_2}{m} \right\}_\Gamma$$

$$\equiv \sum_{\substack{a=0 \\ (a,m)=1}}^{m-1} \bar{\chi}(a N_2) \cdot \pi_E(\gamma_a) \qquad (\mathrm{mod}\ R(E)[\chi]) \ ,$$

(§1.8 and 2.3.2) .

Since $E|\gamma_a = E$, $\pi_E(\gamma_a \beta) = \pi_E(\gamma_a) + \pi_E(\beta)$. Hence

$$\Lambda(\varphi_E, \chi) \equiv \sum_{\substack{a=0 \\ (a,m)=1}}^{m-1} \bar{\chi}(aN_2) \cdot \pi_E(\gamma_a \beta) \qquad (\text{mod } R(E)[\chi]) \quad .$$

Now $\gamma_a \beta = \begin{pmatrix} aN_2 & * \\ m & N_1 a' \end{pmatrix} \in SL_2(\mathbb{Z})$, so by 2.3.3 (b),

$$\pi_E(\gamma_a \beta) = \frac{aN_2}{m} a_0(E) + \frac{N_1 a'}{m} a_0(E|\beta) - e_E\begin{pmatrix} 1 & aN_2 \\ 0 & m \end{pmatrix} \quad .$$

Thus

$$\Lambda(\varphi_E, \chi) \equiv \left(N_2 \cdot \bar{\chi}(N_2) \cdot a_0(E)\right) \cdot \left(\sum_{\substack{a=0 \\ (a,m)=1}}^{m-1} \bar{\chi}(a) \cdot \frac{a}{m}\right)$$

$$+ \left(N_1 \cdot a_0(E|\beta)\right) \cdot \left(\sum_{\substack{a=0 \\ (a,m)=1}}^{m-1} \bar{\chi}(aN_2) \cdot \frac{a'}{m}\right)$$

$$- \sum_{\substack{a=0 \\ (a,m)=1}}^{m-1} \bar{\chi}(aN_2) \cdot e_E\begin{pmatrix} 1 & aN_2 \\ 0 & m \end{pmatrix}$$

$$(\text{mod } R(E)[\chi]) \quad .$$

In each of the first two terms the first factor is in $R(E)$ and the second factor is a generalized Bernoulli number $(\text{mod } \mathbb{Z})$. Applying the lemma to the third term, we obtain

$$\Lambda(\varphi_E, \chi) \equiv \bar{\chi}(N_2) \cdot \mathbf{B}_1(\bar{\chi}) \cdot N_2 a_0(E) + \chi(N_1) \cdot \mathbf{B}_1(\chi) \cdot N_1 a_0(E|\beta)$$

$$+ \frac{\tau(\bar{\chi})}{2\pi i} \cdot L(E, \chi, 1) \qquad (\text{mod } R(E)[\chi]) \quad .$$

□

The main term in this formula is $\frac{1}{2\pi i} \cdot \tau(\bar{\chi}) \cdot L(E, \chi, 1)$. In fact, unless χ or $\bar{\chi}$ is an "exceptional" character (definition 3.1.3), the other two terms are zero, as we shall see in Corollary 3.1.5.

Let $\mathfrak{O}$ be an integral domain of characteristic 0, and suppose χ is $\mathfrak{O}$-valued. Let $\mathfrak{P} \subseteq \mathfrak{O}$ be a maximal ideal with finite residual characteristic $p > 2$.

Definition 3.1.3: The character χ is exceptional at $\mathfrak{P}$ if

a) $p \mid m$,

and b) $\bar{\chi}(x) \equiv x \pmod{\mathfrak{P}}$ for all $x \in \mathbb{Z}$. □

Suppose now $\mathfrak{O} = \mathbb{Z}[\chi]$.

LEMMA 3.1.4: If χ is nonexceptional at $\mathfrak{P}$ then the generalized Bernoulli number $B_1(\chi) \in \mathbb{Q}[\chi]$ is $\mathfrak{P}$-integral.

Proof: Let $x \in \mathbb{Z}$ such that $\bar{\chi}(x) \not\equiv x \pmod{\mathfrak{P}}$. Then $\bar{\chi}(x) - x$ is a $\mathfrak{P}$-unit of $\mathfrak{O}$. For $r \in \mathbb{Q}$ write $((r))$ for the fractional part of r. Computing modulo $\mathfrak{O}$ we have

$$(\bar{\chi}(x) - x) \cdot B_1(\chi) = \sum_{a=0}^{m-1} \chi(a) \cdot \bar{\chi}(x) \cdot ((\frac{a}{m})) - \sum_{a=0}^{m-1} \chi(a) \cdot x \cdot ((\frac{a}{m}))$$

$$\equiv \sum_{a=0}^{m-1} \chi(a) \cdot ((\frac{ax}{m})) - \sum_{a=0}^{m-1} \chi(a) \cdot ((\frac{ax}{m}))$$

$$\equiv 0 \pmod{\mathfrak{O}} .$$ □

For an $\mathfrak{O}$-module M write $M_{\mathfrak{P}}$ for the completion of M at $\mathfrak{P}$. Let $\iota_{\mathfrak{P}} : M \to M_{\mathfrak{P}}$ denote the completion homomorphism.

The following is now a simple corollary of Theorem 3.1.2.

COROLLARY 3.1.5: Suppose the following two conditions are satisfied:

a) either $a_0(E) = 0$ or $\bar{\chi}$ is nonexceptional at $\mathfrak{P}$;

and b) either $a_0(E|\sigma_m \sigma) = 0$ or χ is nonexceptional at $\mathfrak{P}$.

Then

$$\iota_{\mathfrak{P}}(\Lambda(\varphi_E, \chi)) \equiv \iota_{\mathfrak{P}}\left(\frac{\tau(\bar{\chi})}{2\pi i} \cdot L(E, \chi, 1)\right) \pmod{R(E)[\chi]_{\mathfrak{P}}} \quad .$$

□

§3.2. Hecke Operators and Galois Modules:

In 2.4.6 we described an action of $GL_2(\mathbb{A})$ on the space $\mathcal{E}_2$ of weight 2 Eisenstein series of all levels. In particular we obtain an action of $GL_2(\mathbb{Z}/N\mathbb{Z})$ on $\mathcal{E}_2(\Gamma(N))$. For each $d \in (\mathbb{Z}/N\mathbb{Z})^*$ let $\tau(d)$ be the operator associated to the matrix $\begin{pmatrix} 1 & 0 \\ 0 & d \end{pmatrix} \in GL_2(\mathbb{Z}/N\mathbb{Z})$. Then $\tau(d)$ satisfies

$$\phi_{(x,y)}|\tau(d) = \phi_{(x,dy)}$$

for $(x, y) \in \left(\frac{1}{N}\mathbb{Z}/\mathbb{Z}\right)^2 \setminus \{\underline{0}\}$.

Let Γ be a congruence group of type (N_1, N_2) with $N = \mathrm{lcm}(N_1, N_2)$. Then the operators $\tau(d)$ preserve the subspace $\mathcal{E}_2(\Gamma) \subseteq \mathcal{E}_2(\Gamma(N))$.

The action of $\mathrm{Gal}(\mathbb{Q}(\mu_N)/\mathbb{Q})$ on $\underline{\mathrm{cusps}}(\Gamma)$ extends to an action on $\mathrm{Div}^0(\underline{\mathrm{cusps}}(\Gamma);\mathbb{C})$. For $d \in (\mathbb{Z}/N\mathbb{Z})^*$ the element $\tau_d \in \mathrm{Gal}(\mathbb{Q}(\mu_N)/\mathbb{Q})$ is defined by $\tau_d : e^{2\pi i/N} \mapsto e^{2\pi i d/N}$. For $E \in \mathcal{E}_2(\Gamma)$, let $\delta_\Gamma(E) \in \mathrm{Div}^0(\underline{\mathrm{cusps}}(\Gamma);\mathbb{C})$ be the associated residual divisor (§1.8).

PROPOSITION 3.2.1: For $d \in (\mathbb{Z}/N\mathbb{Z})^*$ and $E \in \mathcal{E}_2(\Gamma)$ we have:

(1) $\delta_\Gamma(E|\tau(d)) = \delta_\Gamma(E)^{\tau_d}$,

$\delta_\Gamma(E|\langle d\rangle) = \langle d\rangle^{-1} \cdot \delta_\Gamma(E)$;

(2) $C_{E|\tau(d)} = C_E^{\tau_d}$;

(3) $\quad R(E|\tau(d)) = R(E|\langle d\rangle) = R(E)$;

(4) $\quad \underline{\text{Periods}}(E|\tau(d)) = \underline{\text{Periods}}(E|\langle d\rangle) = \underline{\text{Periods}}(E)$.

Proof: Let $\pi : X(N) \to X(\Gamma)$ be the natural projection. Then π is defined over $\mathbb{Q}$ and

$$\pi^* : \mathrm{Div}^0(\underline{\text{cusps}}(\Gamma)) \longrightarrow \mathrm{Div}^0(\underline{\text{cusps}}(\Gamma(N)))$$

is injective.

For $\underline{x} = \left(\frac{x}{N}, \frac{y}{N}\right) \neq \underline{0}$ we have by Proposition 2.5.1 (b) (ii)

$$\delta_{\Gamma(N)}(\phi_{\underline{x}}) = \frac{N}{2} \sum_{[{a \atop b}]_{\Gamma(N)}} B_2\left(\frac{ax+by}{N}\right) \cdot [{a \atop b}]_{\Gamma(N)} .$$

So
$$\delta_{\Gamma(N)}(\phi_{\underline{x}}|\tau(d)) = \frac{N}{2} \cdot \sum B_2\left(\frac{ax+bdy}{N}\right) \cdot [{a \atop b}]_{\Gamma(N)}$$

$$= \delta_{\Gamma(N)}(\phi_{\underline{x}})^{\tau_d} \qquad \text{(Theorem 1.3.1(b))} .$$

If $E \in \mathcal{E}_2(\Gamma)$ then E is a linear combination of the functions $\phi_{\underline{x}}$, $\underline{x} \in (\frac{1}{N}\mathbb{Z}/\mathbb{Z})^2 \setminus \{\underline{0}\}$. So

$$\pi^*(\delta_\Gamma(E|\tau(d))) = \delta_{\Gamma(N)}(E|\tau(d))$$

$$= \pi^*\left(\delta_\Gamma(E)^{\tau_d}\right) .$$

Since π^* is injective this proves the first part of (1). The second part can be proven in a similar fashion.

The assertions (2) and (3) follow immediately from (1).

Consider the following commutative diagram of exact sequences (1.8.7)

$$\begin{array}{ccccccccc}
0 & \longrightarrow & \underline{\text{Periods}}\,(E)^* & \longrightarrow & R(E)^* & \longrightarrow & C_E & \longrightarrow & 0 \\
 & & & & \| \downarrow & & \downarrow \tau_d & & \\
0 & \longrightarrow & \underline{\text{Periods}}\,(E|\tau(d))^* & \longrightarrow & R(E|\tau(d))^* & \longrightarrow & C_E^{\tau_d} & \longrightarrow & 0
\end{array} .$$

Then $\underline{\text{Periods}}(E)^* = \underline{\text{Periods}}(E|\tau(d))^*$. It now follows that $\underline{\text{Periods}}(E) = \underline{\text{Periods}}(E|\tau(d))$. Similarly we can show $\underline{\text{Periods}}(E) = \underline{\text{Periods}}(E|\langle d\rangle)$. This proves (4). □

PROPOSITION 3.2.2: Let $E \in \mathcal{E}_2(\Gamma)$ be an eigenfunction for the Hecke operators T_ℓ, $\langle \ell \rangle$ for all primes $\ell \nmid N$. Then there is a Dirichlet character $\epsilon_1 : \mathbb{Z} \to \mathbb{C}$ of conductor N for which

$$E|\tau(d) = \epsilon_1(d) \cdot E$$

for all $d \in (\mathbb{Z}/N\mathbb{Z})^*$.

Proof: Let $D(N)$ be the set of Dirichlet characters of conductor N and $\phi(N) = \# D(N)$ be Euler's totient function.

For $\psi \in D(N)$ define the operator $pr(\psi)$ on $\mathcal{E}_2(\Gamma)$ by

$$\mathrm{pr}(\psi) = \frac{1}{\phi(N)} \cdot \sum_{d \in (\mathbb{Z}/N\mathbb{Z})^*} \bar{\psi}(d) \cdot \tau(d) \quad .$$

Then

$$E = \sum_{\psi \in D(N)} E \,|\, \mathrm{pr}(\psi) \quad .$$

For each $\psi \in D(N)$ and $\ell \nmid N$

$$E \,|\, \mathrm{pr}(\psi) \cdot T_\ell = (\psi(\ell) + \ell\epsilon(\ell)\bar{\psi}(\ell)) \cdot E \,|\, \mathrm{pr}(\psi) \quad ,$$

where ϵ is the Nebentypus character of E. On the other hand, since the Hecke operators commute with the $\tau(d)$'s, this eigenvalue must be the same as the eigenvalue of T_ℓ acting on E. We see then that there is exactly one $\psi \in D(N)$ for which $E \,|\, \mathrm{pr}(\psi) \neq 0$, and for this ψ

$$E = E \,|\, \mathrm{pr}(\psi) \quad .$$

Let $\epsilon_1 = \psi$. □

By Proposition 2.4.7 the action of T_ℓ on $\mathcal{E}_2(\Gamma)$ is given by

$$T_\ell = \tau(\ell) + \ell \langle \ell \rangle \tau(\ell)^{-1} \quad .$$

Hence there are Dirichlet characters ϵ_1, ϵ_2 of conductor N such that

(3.2.3)
$$E \,|\, T_\ell = (\epsilon_1(\ell) + \ell \epsilon_2(\ell)) \cdot E \quad ,$$
$$E \,|\, \langle \ell \rangle = \epsilon_1 \epsilon_2(\ell) \cdot E \quad ,$$

for all primes $\ell \nmid N$.

An eigenfunction $E \in \mathcal{E}_2(\Gamma)$ will be said to have signature (ϵ_1, ϵ_2) if 3.2.3 is satisfied for all $\ell \nmid N$.

Let $G = \mathrm{Gal}(\mathbb{Q}(\mu_N)/\mathbb{Q})$ and Δ be the image of $\langle\ \rangle : (\mathbb{Z}/N\mathbb{Z})^* \to \mathbb{T}^*$. Let $\mathbb{Z}[\epsilon_1, \epsilon_2]$ (respectively $\mathbb{Q}[\epsilon_1, \epsilon_2]$) be the ring generated by the values of ϵ_1, ϵ_2 over $\mathbb{Z}$ (respectively over $\mathbb{Q}$).

THEOREM 3.2.4: Let $E \in \mathcal{E}_2(\Gamma; \mathbb{Q}[\epsilon_1, \epsilon_2])$ have signature (ϵ_1, ϵ_2). Then C_E is a cyclic $\mathbb{Z}[G, \Delta]$-module.

The representation

$$\mathbb{Z}[G, \Delta] \longrightarrow \mathrm{End}(C_E)$$

factors through the map $\mathbb{Z}[G, \Delta] \to \mathbb{Z}[\epsilon_1, \epsilon_2]$ defined by $\tau_d \mapsto \epsilon_1(d)$, $\langle \ell \rangle \mapsto \epsilon_1 \epsilon_2(\ell)$.

Proof: We have $0 \neq R(E) \subseteq$ Periods $(E) \subseteq \mathbb{Q}[\epsilon_1, \epsilon_2]$. By 3.2.1 and 3.2.2, $R(E) = R(E|\tau(d)) = \epsilon_1(d) \cdot R(E)$, and similarly $R(E) = \epsilon_1 \epsilon_2(d) \cdot R(E)$. Hence $R(E)$ is a fractional ideal in $Q[\epsilon_1, \epsilon_2]$. A similar argument shows Periods (E) is a fractional ideal. It follows that $A(E)$ is a cyclic $\mathbb{Z}[\epsilon_1, \epsilon_2]$-module. By duality C_E also inherits the structure of a cyclic $\mathbb{Z}[\epsilon_1, \epsilon_2]$-module. For $\alpha \in \mathbb{Z}[\epsilon_1, \epsilon_2]$ let t_α be the associated endomorphism of C_E.

Let $x \in C_E$ and $d \in (\mathbb{Z}/N\mathbb{Z})^*$. Let $\phi \in R(E)^*$ be such that $\phi(\delta(E)) \in \mathrm{Div}^0$(cusps) represents x. Then x^{τ_d} is represented by

$\phi(\delta(E|\tau(d))) = \phi(\epsilon_1(d) \cdot \delta(E))$. Hence $x^{\tau_d} = t_{\epsilon_1(d)}(x)$. Similarly we can show $\langle d \rangle (x) = t_{\epsilon_1 \epsilon_2(d)}(x)$. □

Let $\mathrm{sgn}(E) = -\mathrm{sgn}(\epsilon_1)$, and write $H(Y)$ for $H_1(Y(\Gamma); \mathbb{Z})$.

PROPOSITION 3.2.5: There is a π-homomorphism $\eta_E : H(Y)^{\mathrm{sgn}(E)} \to$ Periods (E) such that the following diagram is commutative:

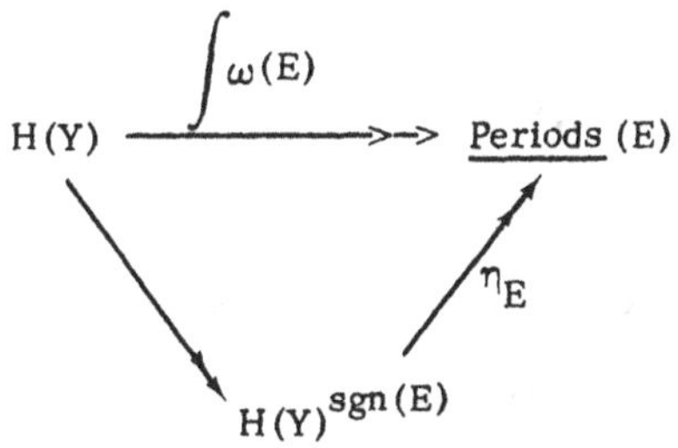

Proof: Let $\gamma \in \Gamma$ and $[\gamma]$ be the associated element of $H(Y)$. In the notation of §2.5

$$\int_{[\gamma]} \omega(E) = \xi_E(\gamma) \quad .$$

Since $\iota_*[\gamma] = [\iota^{-1}\gamma\iota]$, $\iota = \begin{pmatrix} -1 & 0 \\ 0 & 1 \end{pmatrix}$ we have

$$\begin{aligned} \int_{\iota_*[\gamma]} \omega(E) &= \xi_E(\iota^{-1}\gamma\iota) \\ &= \frac{1}{2}\left(\pi_E(\iota^{-1}\gamma\iota) - \pi_{E|\tau(\iota)}(\gamma)\right) \\ &= \mathrm{sgn}(E) \cdot \xi_E(\gamma) \quad . \end{aligned}$$

The last equality follows from Proposition 3.2.2 with $d = -1$. □

§3.3. An Aside on Dirichlet L-functions:

Let m be a positive integer and

$$\psi : \mathbb{Z} \longrightarrow \mathbb{C}$$

be a function on $\mathbb{Z}$ which is periodic with period m. Typically ψ will be a Dirichlet character. For $\mathrm{Re}(s) > 1$ we may define

$$L(\psi, s) \overset{\mathrm{dfn}}{=\!=} \sum_{n=1}^{\infty} \psi(n)\, n^{-s} \quad .$$

As a function of s, $L(\psi, s)$ extends to a meromorphic function on all of $\mathbb{C}$, with a possible simple pole only at $s = 1$.

In this short section we record for easy reference the functional equation and special values (at integers $s \leq 0$) of $L(\psi, s)$. For proofs see ([22], Chapter XIV), (compare also ([18], §§1 and 2)). We also state two simple identities for Gauss sums of Dirichlet characters. These results will be used in the next sections.

Since any function on $\mathbb{Z}$ may be written as a sum of an even and an odd function, we may assume from the outset that ψ is either even or odd. Write $\mathrm{sgn}(\psi) = \pm 1$ for the parity of ψ. Let $\hat{\psi} : \mathbb{Z} \to \mathbb{C}$ be the Fourier transform of ψ defined by

$$\hat{\psi}(n) = \sum_{a=0}^{m-1} \psi(a)\, e^{2\pi i a n/m} \quad .$$

Then $\hat{\psi}$ also has period m and $\mathrm{sgn}(\hat{\psi}) = \mathrm{sgn}(\psi)$.

Let

$$A(\psi) = \sqrt{\frac{m}{\pi}} \quad ,$$

$$\Gamma_\psi(s) = \begin{cases} \Gamma(s/2) & \text{if} \quad \operatorname{sgn}(\psi) = 1 \quad , \\ \Gamma(\frac{1+s}{2}) & \text{if} \quad \operatorname{sgn}(\psi) = -1 \quad ; \end{cases}$$

$$\Phi(\psi, s) = A(\psi)^s \cdot \Gamma_\psi(s) \cdot L(\psi, s) \quad .$$

3.3.1. Functional Equation:

$$\Phi(\psi, s) = \frac{1}{\sqrt{(\operatorname{sgn} \psi)\, m}} \cdot \Phi(\hat{\psi}, 1 - s) \quad ,$$

with the principal branch of the square root. □

3.3.2. Special Values: For $n \geq 1$,

$$L(\psi, 1 - n) = -\frac{B_n(\psi)}{n} \quad ,$$

where

$$B_n(\psi) = m^{n-1} \cdot \sum_{a=1}^{m} \psi(a) \cdot B_n(\tfrac{a}{m})$$

and $B_n(x)$ is the usual Bernoulli polynomial. □

3.3.3. Gauss Sums:

a) Let ψ be a primitive Dirichlet character; then for $n \in \mathbb{Z}$

$$\hat{\psi}(n) = \tau(\psi) \cdot \overline{\psi}(n) \quad .$$

b) Let ψ_1, ψ_2 be two not necessarily primitive Dirichlet characters defined modulo m_1, m_2 respectively. Suppose $(m_1, m_2) = 1$. Then for $n \in \mathbb{Z}$,

$$\hat{\psi}_1(n) \cdot \hat{\psi}_2(n) = \overline{\psi}_1(m_2) \cdot \overline{\psi}_2(m_1) \cdot \widehat{(\psi_1 \psi_2)}(n) \quad .$$

□

§3.4. Eigenfunctions in the Space of Eisenstein Series

Hecke ([15], Werke, S. 690, Satz 44) has shown that the space of Dirichlet series obtained by Mellin transform from the space of weight two Eisenstein series is generated by elements of the form

$$N_2^{s-1} \cdot L(\varepsilon_1, s) \cdot L(\varepsilon_2, s-1)$$

where $\varepsilon_1, \varepsilon_2$ are Dirichlet characters. In this section we introduce a related system of Eisenstein series, and describe the cohomology classes associated to this system by §1.8.

Let $\varepsilon_1, \varepsilon_2 : \mathbb{Z} \to \mathbb{C}$ be two not necessarily primitive Dirichlet characters of conductors N_1, N_2 respectively. We assume not both N_1 and N_2 are equal to 1 and that the character $\varepsilon = \varepsilon_1 \cdot \varepsilon_2$ is even.

<u>Definition 3.4.1</u>: Let $E(\varepsilon_1, \varepsilon_2; z)$ be the Eisenstein series defined by

$$E(\varepsilon_1, \varepsilon_2; z) = \sum_{x=0}^{N_2-1} \sum_{y=0}^{N_1-1} \varepsilon_2(x) \cdot \bar{\varepsilon}_1(y) \cdot \phi_{\left(\frac{x}{N_2}, \frac{y}{N_1}\right)}(z) \quad .$$

□

Let N be the least common multiple of N_1, N_2 and let Γ be the congruence group $\Gamma_1(N_1, N_2)$ of level N.

PROPOSITION 3.4.2: Let $E(z) = E(\varepsilon_1, \varepsilon_2; z)$. Then

a) $E(z)$ is modular of weight two for Γ with Nebentypus character $\varepsilon = \varepsilon_1 \cdot \varepsilon_2$.

b) $L(E, s) = -2 \cdot N_2^{s-1} \cdot L(\hat{\bar{\varepsilon}}_1, s) \cdot L(\varepsilon_2, s-1)$.

c) Let χ be a nontrivial primitive Dirichlet character of conductor m prime to N. Suppose $\operatorname{sgn}(\chi) = \operatorname{sgn}(E)$ $(= -\varepsilon_1(-1))$. Then

$$\left(\frac{\tau(\bar{\chi})}{2\pi i}\right) L(E, \chi, 1) = -\chi(N_1)\bar{\chi}(N_2)\,\varepsilon_1(m) \cdot \mathbf{B}_1(\bar{\varepsilon}_1\bar{\chi}) \cdot \mathbf{B}_1(\varepsilon_2\chi) .$$

<u>Proof</u>: a) Each of the functions $\phi_{(\frac{x}{N_2}, \frac{y}{N_1})}$ is modular for Γ, hence so is E. For a prime ℓ not dividing N, let $\sigma_\ell^{(N)} \in SL_2(\mathbb{Z})$ be such that

$$\sigma_\ell^{(N)} \equiv \begin{pmatrix} \ell' & 0 \\ 0 & \ell \end{pmatrix} \pmod{N}$$

where $\ell\ell' \equiv 1 \pmod{N}$. Then

$$\begin{aligned} E|\langle \ell \rangle &= E|\sigma_\ell^{(N)} \\ &= \sum_{x=0}^{N_2-1} \sum_{y=0}^{N_1-1} \varepsilon_2(x) \cdot \bar{\varepsilon}_1(y) \cdot \phi_{(\frac{\ell' x}{N_2}, \frac{\ell y}{N_1})} \\ &= \varepsilon_1 \cdot \varepsilon_2(\ell) \cdot E \\ &= \varepsilon(\ell) \cdot E . \end{aligned}$$

b) We compute the Fourier expansion of $E(z)$. For

$\underline{x} = (\frac{x}{N_2}, \frac{y}{N_1}) \in (\mathbb{Q}/\mathbb{Z})^2$, define $P_{\underline{x}}(z)$ by

$$P_{\underline{x}}(z) = \sum_{\substack{k \equiv \frac{x}{N_2}(1) \\ k \in \mathbb{Q}^+}} k \cdot \sum_{m=1}^{\infty} q(m(kz + \frac{y}{N_1}))$$

so that

$$\phi_{\underline{x}}(z) = \frac{1}{2} B_2(\frac{x}{N_2}) - P_{\underline{x}}(z) - P_{-\underline{x}}(z) \quad .$$

We have

$$\sum_{x=0}^{N_2-1} \sum_{y=0}^{N_1-1} \epsilon_2(x) \cdot \bar{\epsilon}_1(y) \cdot P_{(\frac{x}{N_2}, \frac{y}{N_1})}$$

$$= \sum_{m=1}^{\infty} \left(\sum_{y=0}^{N_1-1} \bar{\epsilon}_1(y) \cdot q(\frac{my}{N_1}) \right) \cdot \left(\sum_{x=0}^{N_2-1} \sum_{\substack{k \equiv \frac{x}{N_2}(1) \\ k \in \mathbb{Q}^+}} k \cdot \epsilon_2(x) \cdot q(mkz) \right)$$

$$= \frac{1}{N_2} \cdot \sum_{m=1}^{\infty} \hat{\bar{\epsilon}}_1(m) \cdot \sum_{x=0}^{N_2-1} \sum_{\substack{k \equiv x(N_2) \\ k \in \mathbb{Z}^+}} k \cdot \epsilon_2(x) \cdot q(\frac{mkz}{N_2})$$

$$= \frac{1}{N_2} \cdot \sum_{m=1}^{\infty} \hat{\bar{\epsilon}}_1(m) \cdot \sum_{k=1}^{\infty} k \cdot \epsilon_2(k) \cdot q(\frac{mkz}{N_2})$$

$$= \frac{1}{N_2} \cdot \sum_{n=1}^{\infty} \sigma_{\hat{\bar{\epsilon}}_1, \epsilon_2}(n) \cdot q(\frac{nz}{N_2})$$

where

$$\sigma_{\hat{\bar{\epsilon}}_1, \epsilon_2}(n) = \sum_{\substack{d|n \\ d>0}} (\hat{\bar{\epsilon}}_1(d) \cdot \epsilon_2(\tfrac{n}{d}) \cdot \tfrac{n}{d}) \quad .$$

Hence

$$L(E, s) = \frac{-2}{N_2} \cdot \sum_{n=1}^{\infty} \sigma_{\hat{\bar{\epsilon}}_1, \epsilon_2}(n) \cdot (\frac{n}{N_2})^{-s}$$

$$= -2N_2^{s-1} \cdot L(\hat{\bar{\epsilon}}_1, s) \cdot L(\epsilon_2, s-1) \quad .$$

c) We have

$$L(E, \chi, s) = -2\bar{\chi}(N_2) \cdot N_2^{s-1} \cdot L(\epsilon_2\chi, s-1) \cdot L(\hat{\bar{\epsilon}}_1 \cdot \chi, s) \quad .$$

By 3.3.3 a) and b),

$$\hat{\bar{\epsilon}}_1 \cdot \chi = \frac{1}{\tau(\bar{\chi})} \cdot \hat{\bar{\epsilon}}_1 \cdot \hat{\bar{\chi}}$$

$$= \frac{\epsilon_1(m) \cdot \chi(N_1)}{\tau(\bar{\chi})} \cdot \widehat{(\bar{\epsilon}_1 \bar{\chi})} \quad .$$

So

$$L(E, \chi, s) = \frac{-2\bar{\chi}(N_2)\,\chi(N_1)\,\epsilon_1(m)}{\tau(\bar{\chi})} \cdot N_2^{s-1} \cdot L(\epsilon_2\chi, s-1) \cdot L(\widehat{(\bar{\epsilon}_1\bar{\chi})}, s)$$

By 3.3.1 and 3.3.2

$$L(\widehat{(\bar{\epsilon}_1\bar{\chi})}, 1) = \pi i \cdot L(\bar{\epsilon}_1\bar{\chi}, 0)$$

$$= -\pi i \cdot B_1(\bar{\epsilon}_1\bar{\chi}) \quad .$$

Since $L(\epsilon_2\chi, 0) = -B_1(\epsilon_2\chi)$ we arrive at the desired result:

$$L(E,\chi,1) = \frac{-2\pi i\,\chi(N_1)\,\overline{\chi}(N_2)\,\epsilon_1(m)}{\tau(\overline{\chi})} \cdot \mathbf{B}_1(\epsilon_2\chi)\cdot \mathbf{B}_1(\overline{\epsilon}_1\overline{\chi}) \quad .$$

□

We now describe the cohomology class, φ_E, and its special values. We may view φ_E as the homomorphism

$$\varphi_E : \Gamma \longrightarrow A(E)$$

obtained by reducing the image of the homomorphism

$$\xi_E : \Gamma \longrightarrow \mathbb{Q}[\epsilon_1, \epsilon_2]$$

modulo the group, $R(E)$, of residues of $\omega(E) = E(z)\,dz$ on the modular curve $X(\Gamma)$.

Combining the last proposition with 3.1.2 we find the special values of φ_E.

COROLLARY 3.4.3: Let $E = E(\epsilon_1, \epsilon_2)$ and χ be as in 3.4.2 c). Then

$$\begin{aligned}\Lambda(\varphi_E,\chi) \equiv\ & \overline{\chi}(N_2)\,\mathbf{B}_1(\overline{\chi})\cdot N_2\,a_0(E) + \epsilon(m)\,\chi(N_1)\,\mathbf{B}_1(\chi)\cdot N_1\,a_0(E\,|\,\sigma)\\ & - \chi(N_1)\,\overline{\chi}(N_2)\,\epsilon_1(m)\cdot \mathbf{B}_1(\overline{\epsilon}_1\overline{\chi})\,\mathbf{B}_1(\epsilon_2\chi)\\ & \pmod{R(E)[\chi]} \quad .\end{aligned}$$

□

The constant terms $a_0(E)$ and $a_0(E\,|\,\sigma)$ are given in the next proposition. As remarked in 3.1.5 the first two terms of this expression for $\Lambda(\varphi_E, \chi)$ are zero (mod $R(E)[\chi]$) unless χ or $\overline{\chi}$ is in a set of "exceptional" characters (Definition 3.1.3).

The group $R(E)$ can be described as follows. For $a, c \in \mathbb{Z}$ with $(a,c) = 1$, let

$$\mathrm{ind}_\Gamma(\tfrac{a}{c})$$

denote the ramification index of the cusp $\{\frac{a}{c}\}_\Gamma$ on $X(\Gamma)$ over the unique cusp on $X(1)$. Also, let $\gamma_{\{\frac{a}{c}\}} \in SL_2(\mathbb{Z})$ be such that

$$\gamma_{\{\frac{a}{c}\}} \cdot (i\infty) = \frac{a}{c} \quad .$$

Then $R(E)$ is generated by the elements in $\mathbb{P}^1(\mathbb{Q})$ modulo Γ.

$$\mathrm{ind}_\Gamma(\tfrac{a}{c}) \cdot a_0(E|\gamma_{\{\frac{a}{c}\}}) \quad ,$$

where $\frac{a}{c}$ runs through a system of representatives

The ramification index, $\mathrm{ind}_\Gamma(\frac{a}{c})$, is easily found. It is the smallest positive integer, n, such that the matrix

$$\begin{pmatrix} 1 - nac & na^2 \\ -nc^2 & 1 + nac \end{pmatrix}$$

is in Γ.

In Section 2.5 we proved the following identity: for $\gamma = \begin{pmatrix} a & b \\ c & d \end{pmatrix} \in SL_2(\mathbb{Z})$ with $c \geq 0$,

$$\xi_E(\gamma) = \begin{cases} \frac{a}{c} \cdot a_0(E) + \frac{d}{c} \cdot a_0(E|\gamma) - s_E(\frac{a}{c}) & \text{if} \quad c > 0 \;, \\ \frac{b}{d} \cdot a_0(E) & \text{if} \quad c = 0 \;. \end{cases}$$

So to describe the function

$$\xi_E : SL_2(\mathbb{Z}) \longrightarrow \mathbb{Q}[\varepsilon_1, \varepsilon_2]$$

and the group $R(E)$, we only need to know the constant terms $a_0(E|\gamma)$, $\gamma \in SL_2(\mathbb{Z})$, and the function

$$s_E : \mathbb{Q} \longrightarrow \mathbb{Q}[\varepsilon_1, \varepsilon_2] \quad .$$

These are given in

PROPOSITION 3.4.4 : Let $a, c \in \mathbb{Z}$ with $(a,c) = 1$, and $c \geq 0$, then

a) If $c \neq 0$,

$$s_E(\tfrac{a}{c}) = \sum_{\nu=0}^{N_2 c - 1} \varepsilon_2(\nu) \cdot \mathbf{B}_1\left(\frac{\nu}{N_2 c}\right) \cdot \mathbf{B}_{1,\bar{\varepsilon}_1}\left(\frac{N_1 a\nu}{N_2 c}\right) \quad ;$$

b) $$a_0\left(E|\gamma_{\{\frac{a}{c}\}}\right) = \frac{1}{2} \sum_{x=0}^{N_2-1} \sum_{y=0}^{N_1-1} \varepsilon_2(x) \cdot \bar{\varepsilon}_1(y) \cdot \mathbf{B}_2\left(\frac{ax}{N_2} + \frac{cy}{N_1}\right) \quad ;$$

c) In case $\varepsilon_1, \varepsilon_2$ are <u>primitive</u> Dirichlet characters, $R(E)$ is generated as a $\mathbb{Z}[\varepsilon_1, \varepsilon_2]$-module by $\frac{1}{2}\mathbf{B}_2(\varepsilon_1, \varepsilon_2)$ where

$$\mathbf{B}_2(\varepsilon_1, \varepsilon_2) = N \cdot \sum_{x=0}^{N_2-1} \sum_{y=0}^{N_1-1} \varepsilon_2(x) \cdot \bar{\varepsilon}_1(y) \cdot \mathbf{B}_2\left(\frac{x}{N_2} + \frac{y}{N_1}\right) \quad .$$

The residual divisor is given by

$$\delta_\Gamma(E) = \frac{1}{2} \cdot \mathbf{B}_2(\epsilon_1, \epsilon_2) \cdot \sum_{\substack{a \in (\mathbb{Z}/N_2\mathbb{Z})^* \\ c \in (\mathbb{Z}/N_1\mathbb{Z})^*}} \epsilon_1(c) \cdot \bar{\epsilon}_2(a) \cdot \begin{bmatrix} a \\ c \end{bmatrix}_\Gamma .$$

Proof: (a) can be proven by a straightforward calculation using Proposition 2.5.4. To prove (b) and (c) we use the formula $a_0(\phi_{(x_1,x_2)}) = \frac{1}{2}\mathbf{B}_2(x_1)$ for $(x_1, x_2) \in (\mathbb{Q}/\mathbb{Z})^2 \setminus \{\underline{0}\}$. □

§3.5. Nonvanishing Theorems:

The following theorem was proved for prime m by L. Washington [45] and for general m by E. Friedman [7]. Let

$$\mu_{m^\infty} = \{\xi \in \mathbb{C}^* \mid \xi^{m^k} = 1 \text{ for some } k \geq 0\} \quad .$$

THEOREM 3.5.1: Let $\mathfrak{P}$ be a prime of $\overline{\mathbb{Q}}$, and $m > 0$ be an integer prime to $\mathfrak{P}$. Let L be a finite abelian extension of $\mathbb{Q}$. Then the set of odd primitive Dirichlet characters χ of $\mathrm{Gal}(L(\mu_{m^\infty})/\mathbb{Q})$ such that

$$\frac{1}{2} B_1(\chi) \equiv 0 \pmod{\mathfrak{P}}$$

is finite. □

This theorem can sometimes be used to prove nonvanishing results for the values $\Lambda(\varphi_E, \chi)$. The following theorem is an example.

THEOREM 3.5.2: Let ϵ_1, ϵ_2 be primitive Dirichlet characters of conductors N_1, N_2 respectively. Let $N = \mathrm{lcm}(N_1, N_2) > 1$. Let $2 \nmid \mathfrak{P}$ be an odd prime of $\overline{\mathbb{Q}}$ for which $R(E) \subseteq \mathfrak{P}$, and let $m > 0$ be an integer prime to $\mathfrak{P}$. Then the set of primitive Dirichlet characters χ of $\mathrm{Gal}(\mathbb{Q}(\mu_{m^\infty})/\mathbb{Q})$ for which $\mathrm{sgn}(\chi) = \mathrm{sgn}(E)$ and

$$\Lambda(\varphi_E, \chi) = 0$$

is finite.

<u>Proof</u>: This is immediate from Theorem 3.5.1 and Corollary 3.4.3. □

§3.6. The Group of Periods:

Let Γ be a group of type (N_1, N_2) and $\varepsilon_1, \varepsilon_2$ be not necessarily primitive Dirichlet characters of conductors N_1, N_2 respectively such that $E = E(\varepsilon_1, \varepsilon_2)$ is modular for Γ. As always, we assume $N = \operatorname{lcm}(N_1, N_2) > 1$. The most difficult task in the determination of the group $C_E \subseteq \operatorname{Pic}^0(X(\Gamma))$ is to find the group Periods (E) of periods on $Y(\Gamma)$. In this section we prove a partial result which is useful in special examples (see §4.3).

The Eisenstein series E has signature $(\varepsilon_1, \varepsilon_2)$. In §3.2 we showed that the groups Periods (E), R(E) are fractional ideals in $\mathbb{Q}[\varepsilon_1, \varepsilon_2]$. For the sake of simplicity we will localize these modules away from 2. Let $\mathfrak{O} = \mathbb{Z}[\frac{1}{2}, \varepsilon_1, \varepsilon_2]$.

Let

$$S(\varepsilon_i) = \sum_{\nu=0}^{N_i - 1} \varepsilon_i(\nu) .$$

So

$$S(\varepsilon_i) = \begin{cases} \phi(N_i) & \text{if} \quad \varepsilon_i = 1_{N_i} , \\ 0 & \text{otherwise} , \end{cases}$$

where ϕ denotes Euler's totient function.

Let $\mathfrak{a} = \mathfrak{O} + \mathfrak{O} \cdot R(E)$ and $\mathfrak{b} \subseteq \mathbb{Q}[\varepsilon_1, \varepsilon_2]$ be the $\mathfrak{O}$-module generated by the set $\{1, \mathbf{B}_1(\bar{\varepsilon}_1), \mathbf{B}_1(\varepsilon_2), S(\varepsilon_1)\,\mathbf{B}_2(\varepsilon_2), S(\varepsilon_2)\,\mathbf{B}_2(\bar{\varepsilon}_1)\}$.

THEOREM 3.6.1: The following inclusions hold:

$$\mathfrak{a} \subseteq \mathfrak{O} \cdot \underline{\text{Periods}}(E) \subseteq \mathfrak{b} \quad .$$

We will need the following lemma.

LEMMA 3.6.2: Let $N > 0$ and ψ be a Dirichlet character of conductor N. Then for $x \in \mathbb{R}$, $0 \leq x \leq N$ we have

$$B_{1,\psi}(x) = B_1(\psi) + \frac{x}{N} \cdot S(\psi) - \psi(-1) \sum_{0 \leq \nu \leq x}^{*} \psi(\nu)$$

where $\sum^*$ denotes the sum with the terms corresponding to the endpoints $0, x$ weighted by a factor of $1/2$.

Proof: We have

$$B_1\left(\frac{x+\nu}{N}\right) - B_1\left(\frac{\nu}{N}\right) = \begin{cases} \frac{x}{N} & \text{if} \quad 0 < \nu < N - x \ , \\ \frac{x}{N} - 1 & \text{if} \quad N - x < \nu < N \ , \\ \frac{x}{N} - \frac{1}{2} & \text{if} \quad \nu = 0, N - x, N \quad . \end{cases}$$

So

$$\begin{aligned} B_{1,\psi}(x) - B_1(\psi) &= \sum_{\nu=1}^{N} \psi(\nu) \cdot \left(B_1\left(\frac{x+\nu}{N}\right) - B_1\left(\frac{\nu}{N}\right)\right) \\ &= \sum_{\nu=1}^{N} \psi(\nu) \cdot \frac{x}{N} - \sum_{N-x \leq \nu \leq N}^{*} \psi(\nu) \\ &= S(\psi) \cdot \frac{x}{N} - \psi(-1) \cdot \sum_{0 \leq \nu \leq x}^{*} \psi(\nu) \quad . \end{aligned}$$

$\square$

Proof of Theorem 3.6.1: We begin by showing $\mathfrak{G} \cdot \underline{\text{Periods}}(E) \subseteq \mathfrak{b}$. Let

$\gamma = \begin{pmatrix} a & N_2 b \\ N_1 c & d \end{pmatrix} \in \Gamma$. We must show $\xi_E(\gamma) \in \mathfrak{b}$. We will show

$(N_1 N_2 c)^2 \cdot \xi_E(\gamma) \in \mathfrak{b}$ and $d^2 \cdot \xi_E(\gamma) \in \mathfrak{b}$. Since $(N_1 N_2 c, d) = 1$ the desired result follows.

By Proposition 3.4.4 we have

$$a_0(E) = \frac{1}{2N_2} \cdot S(\epsilon_1) \cdot \mathbf{B}_2(\epsilon_2)$$

$$a_0(E|\sigma) = \frac{1}{2N_1} \cdot S(\epsilon_2) \cdot \mathbf{B}_2(\bar{\epsilon}_1)$$

where $\sigma = \begin{pmatrix} 0 & -1 \\ 1 & 0 \end{pmatrix}$. By 3.4.4 a)

$$(N_1 N_2 c)^2 \cdot s_E\left(\frac{a}{N_1 c}\right) \in \mathfrak{G} .$$

Since

$$\xi_E(\gamma) = \frac{a}{N_1 c} \cdot a_0(E) + \frac{d}{N_1 c} \cdot a_0(E) - s_E\left(\frac{a}{N_1 c}\right)$$

we have $(N_1 N_2 c)^2 \xi_E(\gamma) \in \mathfrak{b}$.

On the other hand, $\xi_E(\gamma) = \xi_E(\gamma\sigma) - \xi_E(\sigma)$ where

$$\xi_E(\gamma\sigma) = \frac{N_2 b}{d} a_0(E) - \frac{N_1 c}{d} a_0(E|\sigma) - s_E\left(\frac{N_2 b}{d}\right) ,$$

and $$\xi_E(\sigma) = -s_E(0) = -\mathbf{B}_1(\bar{\epsilon}_1) \cdot \mathbf{B}_1(\epsilon_2) .$$

A direct calculation using 3.4.4 shows

$$s_E\left(\frac{N_2 b}{d}\right) = \epsilon_2(d) \cdot \sum_{\nu=0}^{d-1} B_{1,\bar{\epsilon}_1}\left(\frac{N_1 N_2 b\nu}{d}\right) \cdot B_{1,\epsilon_2}\left(\frac{N_2 \nu}{d}\right) .$$

Using the last lemma we find

$$d^2 \cdot s_E\left(\frac{N_2 b}{d}\right) \equiv d^3 \epsilon_2(d) \cdot \mathbf{B}_1(\bar{\epsilon}_1) \cdot \mathbf{B}_1(\epsilon_2) \quad (\text{modulo } \mathfrak{b}) .$$

So

$$d^2 \cdot \xi_E(\gamma) \equiv d^2(1 - d\epsilon_2(d)) \cdot \mathbf{B}_1(\bar{\epsilon}_1) \cdot \mathbf{B}_1(\epsilon_2) \quad (\text{modulo } \mathfrak{b}) .$$

But Lemma 3.1.4 shows $(1 - d\epsilon_2(d)) \cdot \mathbf{B}_1(\epsilon_2) \in \mathfrak{O}$. Hence $d^2 \xi_E(\gamma) \in \mathfrak{b}$.

Next we show $\mathfrak{a} \subseteq \mathfrak{O} \cdot$ <u>Periods</u> (E). Since $R(E) \subseteq$ <u>Periods</u> (E) we only need to show $1 \in \mathfrak{O} \cdot$ <u>Periods</u> (E). Suppose this is not the case. Then there is an odd prime $\mathfrak{P}$ for which <u>Periods</u> (E) $\subseteq \mathfrak{P}$. Let $m > 0$ be prime to $N \cdot \mathfrak{P}$. Corollary 3.1.5 and 3.4.2 c) show that

$$\mathbf{B}_1(\bar{\epsilon}_1 \bar{\chi}) \cdot \mathbf{B}_1(\epsilon_2 \chi) \equiv 0 \qquad (\text{mod } \mathfrak{P})$$

for every primitive $\chi \neq 1$ of conductor dividing m^∞. This contradicts Theorem 3.5.1. □

Chapter 4. Congruences

We would like to translate the congruences of Chapter 3, which are satisfied by the universal special values, into congruences satisfied by the algebraic part of the values $L(f, \chi, 1)$ where f is a parabolic eigenform. We do this modulo certain Eisenstein primes $\mathfrak{P} \subseteq \mathcal{O}(f)$ associated to a pair E, f of eigenfunctions $E \in \mathcal{E}_2(f)$, $f \in \mathcal{S}_2(\Gamma)$ (see §§4.1, 4.2). It is not unreasonable to hope that such congruences will hold for every Eisenstein prime. In fact, if $\Gamma = \Gamma_0(N)$, N prime Mazur ([26], Sec. 7) has proven this. Unfortunately his proof does not generalize and we must make two technical assumptions (4.2.2):

(1) $\dim_k A_f[\mathfrak{P}] = 2$, $(k = \mathcal{O}(f)/\mathfrak{P})$;

(2) $C_E \cap A_f[\mathfrak{P}] \neq 0$.

We define the Eisenstein ideal $I(E)$ in §4.1 and prove that it annihilates the group C_E. If condition (2) is satisfied we show that $\mathfrak{P}$ is an Eisenstein prime associated to the pair E, f. Proposition 4.1.4 identifies the Eisenstein primes as precisely those primes modulo which the Hecke eigenvalues of f are congruent to the eigenvalues of a Galois conjugate E_σ of E. In Chapter 5 we will need to know that an Eisenstein prime is "ordinary" if its residual characteristic is prime to the level. This is proved in Theorem 4.1.6.

Theorem 4.2.3 proves that the algebraic parts of the values $L(f, \chi, 1)$ and $L(E_\sigma, \chi, 1)$ are congruent modulo every odd Eisenstein prime $\mathfrak{P}$ satisfying conditions (1) and (2).

The chapter concludes by applying Theorem 4.2.3 to give two concrete examples. In the first example, f is one of the two newforms for $\Gamma_1(13)$. In the second example, f is the unique newform for $\Gamma_0(7, 7)$.

§4. 1. Eisenstein Ideals:

Let $N_1, N_2 > 0$ and $N = \mathrm{lcm}(N_1, N_2)$ be positive integers and Γ be a congruence group of type (N_1, N_2). Let $\varepsilon_1, \varepsilon_2$ be two not necessarily primitive Dirichlet characters of conductors N_1, N_2 respectively. Fix an Eisenstein series $0 \neq E \in \mathcal{E}_2(\Gamma; \mathbb{Q}[\varepsilon_1, \varepsilon_2])$ of signature $(\varepsilon_1, \varepsilon_2)$ which is an eigenfunction for the Hecke operators T_p, all primes p.

Let $h_E : \mathfrak{J} \to \mathbb{Z}[\varepsilon_1, \varepsilon_2]$ be the homomorphism defined by the equation

$$E|\alpha = h_E(\alpha) \cdot E$$

for $\alpha \in \mathfrak{J}$. Let $\mathcal{J}(E) = \ker(h_E)$.

Definition 4. 1. 1:

(i) The Eisenstein ideal associated to E is the image $I(E) \subseteq \mathbb{T}$ of $\mathcal{J}(E)$ under the natural map $\mathfrak{J} \twoheadrightarrow \mathbb{T}$.

(ii) A prime ideal $P \subseteq \mathbb{T}$ will be called an Eisenstein prime associated to E if $I(E) \subseteq P$. □

THEOREM 4. 1. 2: $I(E) \subseteq \mathrm{Ann}(C_E)$.

Proof: Let $\alpha \in I(E)$ and $x \in C_E$. Let $\beta \in \mathcal{J}(E)$ be an element of the inverse image of α under the map $\mathfrak{J} \twoheadrightarrow \mathbb{T}$ and choose $\phi \in R(E)^*$ so that the divisor class x is represented by $\phi(\delta(E)) \in \mathrm{Div}^0(\underline{\mathrm{cusps}})$. Then

$$\begin{aligned}\alpha \cdot x &= \alpha \cdot \mathrm{cl}(\phi(\delta(E)))\\ &= \mathrm{cl}({}^t\beta \cdot \phi(\delta(E)))\\ &= \mathrm{cl}(\phi(\delta(E|\beta)))\\ &= 0 \quad . \qquad \square\end{aligned}$$

Let $f \in S_2(\Gamma)$ be a parabolic eigenform for Γ. Let $h_f : \mathbb{T} \to \mathcal{O}(f)$ be the homomorphism defined in §1.4. A prime ideal $\mathfrak{P} \subseteq \mathcal{O}(f)$ will be called an Eisenstein prime associated to the pair E, f if $h_f^{-1}(\mathfrak{P})$ is an Eisenstein prime associated to E.

Let $A_f \subseteq \mathrm{Pic}^0(X)$ be the abelian subvariety associated to f ([42], Theorem 7.14). Then $\mathcal{O}(f)$ may be identified with a subring of $\mathrm{End}(A_f/\mathbb{Q})$. For an ideal $\mathfrak{a} \subseteq \mathcal{O}(f)$ we let $A_f[\mathfrak{a}] = \{x \in A_f(\overline{\mathbb{Q}}) \,|\, \alpha \cdot x = 0$ for all $\alpha \in \mathfrak{a}\}$ denote the $\mathfrak{a}$-torsion subgroup of A_f.

COROLLARY 4.1.3: Let $\mathfrak{P} \subseteq \mathcal{O}(f)$ be a prime ideal for which $C_E \cap A_f[\mathfrak{P}] \neq 0$. Then $\mathfrak{P}$ is an Eisenstein prime associated to E, f.

<u>Proof</u>: We have

$$\begin{aligned} I(E) \subseteq \mathrm{Ann}_{\mathbb{T}}(C_E) &\subseteq \mathrm{Ann}_{\mathbb{T}}(C_E \cap A_f[\mathfrak{P}])\\ &= h_f^{-1}(\mathrm{Ann}_{\mathcal{O}(f)}(C_E \cap A_f[\mathfrak{P}]))\\ &= h_f^{-1}(\mathfrak{P}) \quad . \qquad \square\end{aligned}$$

<u>Questions</u>:

(1) Can the inclusion of Theorem 4.1.2 be strengthened to an equality?

(2) Does the converse of Corollary 4.1.3 hold?

An affirmative answer to the first question would imply an affirmative answer to the second one.

In the case $\Gamma = \Gamma_0(N)$, N prime, a result of B. Mazur ([28], II 9.7, 11.1) shows that the answer to question 1 is yes.

In the next proposition we show that the Eisenstein primes are the primes for which the Fourier coefficients of f are congruent modulo $\mathfrak{P}$ to the Fourier coefficients of a Galois conjugate E_σ of E.

By §2.4 E can be expressed as a linear combination

$$E = \sum_{\underline{x} \in (\frac{1}{N}\mathbb{Z}/\mathbb{Z})^2} a_{\underline{x}} \cdot \phi_{\underline{x}}$$

with coefficients $a_{\underline{x}} \in \mathbb{Q}[\varepsilon_1, \varepsilon_2]$. For $\sigma \in \mathrm{Gal}(\overline{\mathbb{Q}}/\mathbb{Q})$ let

$$E_\sigma = \sum_{\underline{x} \in (\frac{1}{N}\mathbb{Z}/\mathbb{Z})^2} a_{\underline{x}}^\sigma \cdot \phi_{\underline{x}} .$$

Since each $\phi_{\underline{x}}$ has rational periods we have for $\gamma \in H_1(Y;\mathbb{Z})$

$$\int_\gamma \omega(E_\sigma) = \left(\int_\gamma \omega(E)\right)^\sigma .$$

We also have $h_{E_\sigma} = \sigma \circ h_E$. To see this let $\gamma \in H_1(Y;\mathbb{Z})$ such that $\int_\gamma \omega(E) \neq 0$. Let $\alpha \in \mathfrak{J}$ and $\beta \in \mathbb{Z}[GL_2^+(\mathbb{Q})]$ represent α. Then β preserves $\mathcal{E}_2(\mathbb{Q})$. We therefore have

$$\sigma \circ h_E(\alpha) \cdot \int_\gamma \omega(E_\sigma) = \left(h_E(\alpha) \cdot \int_\gamma \omega(E)\right)^\sigma$$

$$= \left(\int_\gamma \omega(E|\beta)\right)^\sigma = \left(\sum_{\underline{x}} a_{\underline{x}} \int_\gamma \omega(\phi_{\underline{x}}|\beta)\right)^\sigma$$

$$= \sum_{\underline{x}} a_{\underline{x}}^\sigma \int_\gamma \omega(\phi_{\underline{x}}|\beta) = \int_\gamma \omega(E_\sigma|\beta)$$

$$= h_{E_\sigma}(\alpha) \cdot \int_\gamma \omega(E_\sigma) \quad .$$

Let

$$E(z) = a_0(E) + \sum_{n=1}^{\infty} a_n(E)\, e^{2\pi i n z/N_2} \quad ,$$

and

$$f(z) = \sum_{n=1}^{\infty} a_n(f)\, e^{2\pi i n z/N_2}$$

be the Fourier expansions of E and f. Suppose E and f are normalized so that $a_1(E) = a_1(f) = 1$. Let $\epsilon : (\mathbb{Z}/N\mathbb{Z})^* \to \mathbb{C}^*$ be the Nebentypus character of f. The coefficients $a_n(E)$, $a_n(f)$ are algebraic integers.

PROPOSITION 4.1.4 : Let $\wp \subseteq \mathcal{O}(f)$ be a prime ideal. Then $\wp$ is an Eisenstein prime associated to E, f if and only if there is a prime $\mathfrak{P}$ of $\bar{\mathbb{Q}}$ extending $\wp$ and an element $\sigma \in \mathrm{Gal}(\bar{\mathbb{Q}}/\mathbb{Q})$ such that

(1) $a_n(E_\sigma) \equiv a_n(f) \pmod{\mathfrak{P}}$

for all $n > 0$

and (2) $(\varepsilon_1 \varepsilon_2(d))^\sigma \equiv \varepsilon(d) \pmod{\mathfrak{P}}$

for all $d \in (\mathbb{Z}/N\mathbb{Z})^*$.

Proof: Let $\mathfrak{P}$ be any prime of $\bar{\mathbb{Q}}$ extending $\wp$. Let $\bar{\mathbb{Z}}$ denote the integers of $\bar{\mathbb{Q}}$. Let $k = \mathcal{O}(f)/\wp$ and $\bar{k}$ be an algebraic closure of k. There is a homomorphism

$$\theta : \bar{\mathbb{Z}} \longrightarrow \bar{k}$$

such that the restriction of θ to $\mathcal{O}(f)$ is reduction modulo $\wp$ and $\mathrm{Ker}(\theta) = \mathfrak{P}$.

From the definitions we have $\wp$ is an Eisenstein prime iff there is a homomorphism $\theta_1 : \bar{\mathbb{Z}} \to \bar{k}$ such that the diagram

$$\begin{array}{ccccccc} \mathfrak{J} & \xrightarrow{h_E} & \mathbb{Z}[\varepsilon_1, \varepsilon_2] & \hookrightarrow & \bar{\mathbb{Z}} & \xrightarrow{\theta_1} & \bar{k} \\ \downarrow & & & & & & \| \\ \mathbb{T} & \xrightarrow{h_f} & \mathcal{O}(f) & \hookrightarrow & \bar{\mathbb{Z}} & \xrightarrow{\theta} & \bar{k} \end{array}$$

is commutative.

If $\theta_1 : \bar{\mathbb{Z}} \to \bar{k}$ is such a homomorphism then conditions (1) and (2) are satisfied by letting $\sigma \in \mathrm{Gal}(\overline{\mathbb{Q}}/\mathbb{Q})$ be any element for which $\theta_1 = \theta \circ \sigma$.

Conversely, if (1) and (2) are satisfied let $\theta_1 = \theta \circ \sigma$. □

We conclude this section by showing that an Eisenstein prime $P \subseteq \mathbb{T}$ is "ordinary".

Let $P \subseteq \mathbb{T}$ be a maximal ideal of residual characteristic p prime to N. Let $t_p \in \mathbb{T}$ be the image of the abstract Hecke operator $T_p \in \mathfrak{J}$.

Definition 4.1.5: P is called ordinary if $t_p \notin P$. □

THEOREM 4.1.6: If $P \subseteq \mathbb{T}$ is an Eisenstein prime associated to E of residual characteristic p prime to N. Then P is ordinary.

Proof: Let $\mathfrak{O}(E) = \mathrm{Im}(h_E) \subseteq \mathbb{Z}[\epsilon_1, \epsilon_2]$. Then $\mathbb{Z}[\epsilon_1, \epsilon_2]$ is the integral closure of $\mathfrak{O}(E)$ in its quotient field. Now $h_E(T_p) = \epsilon_1(p) + p\epsilon_2(p)$ is a unit (modulo p) in $\mathbb{Z}[\epsilon_1, \epsilon_2]$ and is therefore also a unit (modulo p) in $\mathfrak{O}(E)$ (by "going up"). Hence

$$\mathfrak{J} = T_p \cdot \mathfrak{J} + p \cdot \mathfrak{J} + \mathcal{J}(E) \quad .$$

Looking at the image in $\mathbb{T}$ we find

$$\mathbb{T} = t_p \cdot \mathbb{T} + p \cdot \mathbb{T} + I(E) \quad .$$

Since $P \supseteq I(E) + p \cdot \mathbb{T}$ we have $t_p \notin P$. □

§4.2. Congruences Satisfied by Values of L-functions:

We preserve the notations of §4.1.

PROPOSITION 4.2.1: Let $\mathfrak{P} \subseteq \mathcal{O}(f)$ be a prime ideal and $k = \mathcal{O}(f)/\mathfrak{P}$. Suppose $2 \notin \mathfrak{P}$. Then the following are equivalent:

(a) $\dim_k A_f[\mathfrak{P}] = 2$;

(b) $H(f)_{\mathfrak{P}}$ is a free rank 2 $\mathcal{O}(f)_{\mathfrak{P}}$-module ;

(c) $H(f)^{\pm}_{\mathfrak{P}}$ are free rank 1 $\mathcal{O}(f)_{\mathfrak{P}}$-modules .

Proof: Let $\mathbb{C}^*_1 = \{\xi \in \mathbb{C}^* \mid |\xi| = 1\}$. Pontrjagin duality gives a perfect pairing

$$A_f \times H(f) \longrightarrow \mathbb{C}^*_1 .$$

The action of $\mathbb{T}$ on A_f is dual to the action on $H(f)$. Therefore if p is the residual characteristic of $\mathfrak{P}$ we have a perfect pairing

$$A_f[\mathfrak{P}] \times H(f)/\mathfrak{P}H(f) \longrightarrow \mu_p .$$

Hence $\dim_k(A_f[\mathfrak{P}]) = \dim_k(H(f)/\mathfrak{P}H(f))$.

Since $p \neq 2$ we have $H(f)_{\mathfrak{P}} \cong H(f)^+_{\mathfrak{P}} \oplus H(f)^-_{\mathfrak{P}}$. So the implications (c) $\Rightarrow$ (b) $\Rightarrow$ (a) are obvious.

Assume condition (a) holds. We will prove (c). Let $h^{\pm} = \dim_k(H(f)^{\pm}/\mathfrak{P}H(f)^{\pm})$. Then $h^+ + h^- = 2$. Since $H(f)^{\pm} \otimes \mathbb{Q}$ are one dimensional $K(f)$ vector spaces we have $H(f)^{\pm}_{\mathfrak{P}} \neq 0$. Hence, Nakayama's lemma tells us $h^+, h^- > 0$. This can only be true if $h^+ = h^- = 1$. Let $\gamma^\pm \in H(f)^\pm_{\mathfrak{p}}$ represent a generator of $H(f)^\pm/\mathfrak{p}H(f)^\pm$. Another application of Nakayama's lemma shows $\gamma^\pm$ generates $H(f)^\pm_{\mathfrak{p}}$ as an $\mathcal{O}(f)_{\mathfrak{p}}$-module. This proves (c). $\square$

Let $\mathfrak{p} \subseteq \mathcal{O}(f)$ be a prime ideal of residual characteristic $p \neq 2$ satisfying the following conditions:

1) the conditions of Proposition 4.2.1;

(4.2.2)

2) $C_E \cap A_f[\mathfrak{p}] \neq 0$.

By Corollary 4.1.3 $\mathfrak{p}$ is an Eisenstein prime associated to E, f.

Let $B = C_E \cap A_f[\mathfrak{p}]$ and $A = B^\wedge$. Also let $H = H_1(X;\mathbb{Z})$ and $H(Y) = H_1(Y;\mathbb{Z})$. We have a commutative diagram

$$\begin{array}{ccccccc}
H(Y)^{\mathrm{sgn}(E)} & \twoheadrightarrow & H^{\mathrm{sgn}(E)} & \twoheadrightarrow & H(f)^{\mathrm{sgn}(E)} & \xrightarrow{\sim} & \underline{\mathrm{Periods}}\,(f)^{\mathrm{sgn}(E)} \\
\downarrow{\scriptstyle \eta_E} & & \downarrow{\scriptstyle \varphi_E^{\mathrm{sgn}(E)}} & & \downarrow & \swarrow & \\
\underline{\mathrm{Periods}}\,(E) & \twoheadrightarrow & A(E) & \twoheadrightarrow & A & &
\end{array} .$$

The commutativity of the left-hand square follows from the definition of $A(E)$ (§1.8) and Proposition 3.2.5. The commutativity of the right-hand square follows from the remarks at the end of §1.7.

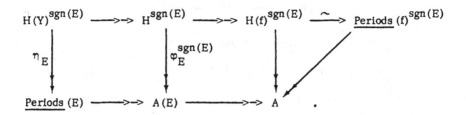

Since the group B is both Hecke invariant and $\text{Gal}(\overline{\mathbb{Q}}/\mathbb{Q})$-invariant, A inherits both a structure of k-module and of $\mathbb{Z}[\epsilon_1, \epsilon_2]$-module. By our assumption 2), $A \neq 0$. By Theorem 3.2.4 A is a cyclic $\mathbb{Z}[\epsilon_1, \epsilon_2]$-module. Let $\underline{a} \in A$ be a generator. By assumption 1) $\underline{a}$ generates A as a k-vector space.

Let $\Omega_E \in \underline{\text{Periods}}(E)$ and $\Omega_f^{\text{sgn}(E)} \in \underline{\text{Periods}}(f)^{\text{sgn}(E)}$ be elements which project to $\underline{a}$ in the above diagram.

Let χ be a nontrivial primitive Dirichlet character of conductor prime to N. Suppose $\text{sgn}(\chi) = \text{sgn}(E)$. Define

$$\Lambda_E(\chi) = \frac{\tau(\overline{\chi})\, L(E, \chi, 1)}{2\pi i\, \Omega_E} \quad,$$

and

$$\Lambda_f(\chi) = \frac{\tau(\overline{\chi})\, L(f, \chi, 1)}{2\pi i\, \Omega_f^{\text{sgn}(E)}} \quad.$$

If $\sigma \in \text{Gal}(\overline{\mathbb{Q}}/\mathbb{Q})$ let $\Omega_{E_\sigma} = \Omega_E^\sigma$. Then Ω_{E_σ} projects to $\underline{a} \in A$ under the natural map $\underline{\text{Periods}}(E_\sigma) \longrightarrow\!\!\!\!\!\rightarrow A$. Define

$$\Lambda_{E_\sigma}(\chi) = \frac{\tau(\overline{\chi})\, L(E_\sigma, \chi, 1)}{2\pi i\, \Omega_{E_\sigma}} \quad.$$

THEOREM 4.2.3: Assume (4.2.2). Let $\mathfrak{P}$ be a prime of $\overline{\mathbb{Q}}$ extending $\mathcal{P}$. Then there is a $\sigma \in \text{Gal}(\overline{\mathbb{Q}}/\mathbb{Q})$ such that $h_{E_\sigma} \equiv h_f$ (modulo $\mathfrak{P}$) and for every nontrivial primitive Dirichlet character χ which is nonexceptional at $\mathfrak{P}$, has conductor prime to N, and $\text{sgn}(\chi) = \text{sgn}(E)$:

1) $\Lambda_f(\chi)$, $\Lambda_{E_\sigma}(\chi)$ are $\mathfrak{P}$-integral

and 2) $\Lambda_f(\chi) \equiv \Lambda_{E_\sigma}(\chi)$ (modulo $\mathfrak{P}$).

Proof: Since A is a one-dimensional k-vector space we have $\text{End}_k(A) \cong k$. Let

$$\theta_E : \mathbb{Z}[\epsilon_1, \epsilon_2] \longrightarrow k = \text{End}_k(A) ,$$

$$\theta : \mathcal{O}(f) \longrightarrow\!\!\!\!\!\rightarrow k$$

be the natural ring homomorphisms. Since A is a cyclic $\mathbb{Z}[\epsilon_1, \epsilon_2]$-module, θ_E is surjective. Let $q = \text{Ker}(\theta_E)$.

For an integral domain R and a prime ideal $P \subseteq R$ write $R_{(P)}$ for the localization of R at P viewed as a subring of the quotient field of R. If M is an R-module let $M_{(P)} = M \otimes_R R_{(P)}$.

Since Periods (E) is a fractional ideal in $\mathbb{Q}[\epsilon_1, \epsilon_2]$, Periods $(E)_{(q)}$ is a cyclic $\mathbb{Z}[\epsilon_1, \epsilon_2]_{(q)}$-module. The $\mathbb{Z}[\epsilon_1, \epsilon_2]$-module homomorphism Periods (E) $\to A$ sends Ω_E to the generator $\underline{a}$. Hence

$$\Omega_E^{-1} \cdot \text{Periods}(E) \subseteq \mathbb{Z}[\epsilon_1, \epsilon_2]_{(q)} ,$$

and

$$\Omega_E^{-1} \cdot R(E) \subseteq q \cdot \mathbb{Z}[\epsilon_1, \epsilon_2]_{(q)} .$$

The $K(f)$-vector space $H(f)^{\text{sgn}(E)} \otimes \mathbb{Q}$ is one-dimensional, so by assumption 1),

$$\Omega_f^{-1} \cdot \underline{\text{Periods}}\,(f)^{\mathrm{sgn}(E)} \subseteq \mathcal{O}(f)_{(\mathfrak{P})} \quad .$$

We have the following commutative diagram:

$$\begin{array}{ccccc}
 & & \overset{\psi_E}{\frown\!\!\!\longrightarrow} & & \\
H(Y) & \xrightarrow{\int \omega(E)} & \underline{\text{Periods}}\,(E) & \xrightarrow{\Omega_E^{-1}} & \mathbb{Z}[\epsilon_1, \epsilon_2]_{(q)} \\
S \downarrow & & \downarrow & & \downarrow \theta_E \\
 & & A & \xrightarrow{\underline{a}^{-1}} & k \\
 & & \uparrow & & \uparrow \theta \\
H & \xrightarrow{\varphi_f^{\mathrm{sgn}(E)}} & \underline{\text{Periods}}\,(f)^{\mathrm{sgn}(E)} & \xrightarrow{(\Omega_f^{\mathrm{sgn}(E)})^{-1}} & \mathcal{O}(f)_{(\mathfrak{P})} \\
 & & \underset{\psi_f}{\smile\!\!\!\longrightarrow} & &
\end{array} \quad .$$

The map S is the natural one and ψ_E, ψ_f are defined by the commutativity.

Let $\bar{k}$ be an algebraic closure of k and extend θ to a homomorphism $\theta : \bar{\mathbb{Z}} \to \bar{k}$ with kernel $\mathfrak{B}$. Extend θ_E to a homomorphism $\theta_E : \bar{\mathbb{Z}} \to \bar{k}$. Let $\mathfrak{D} = \ker(\theta_E)$ and $\sigma \in \mathrm{Gal}(\bar{\mathbb{Q}}/\mathbb{Q})$ such that $\theta \circ \sigma = \theta_E$.

Let $\psi_{E_\sigma} = \sigma \circ \psi_E$. Then ψ_{E_σ} has values in $\bar{\mathbb{Z}}_{(\mathfrak{B})}$. The following diagram is commutative

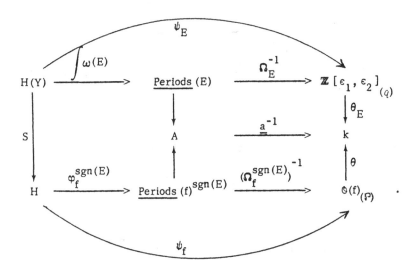

$$\begin{array}{ccc}
H(Y) & \xrightarrow{\psi_{E_\sigma}} & \overline{\mathbb{Z}}_{(\mathfrak{P})} \\
 & & \big\downarrow \theta \\
S \big\downarrow & & \bar{k} \\
 & & \big\uparrow \theta \\
H & \xrightarrow[\psi_f]{} & \overline{\mathbb{Z}}_{(\mathfrak{P})}
\end{array} \quad .$$

Let $(\psi_{E_\sigma} \otimes 1) : H(Y) \otimes \overline{\mathbb{Z}} \to \overline{\mathbb{Z}}_{(\mathfrak{P})}$ and $(\psi_f \otimes 1) : H \otimes \overline{\mathbb{Z}} \to \overline{\mathbb{Z}}_{(\mathfrak{P})}$.

Let χ be as in the theorem. Let $\Lambda(\chi) \in H \otimes \overline{\mathbb{Z}}$ be the universal special value, and $\widetilde{\Lambda(\chi)} \in H(Y) \otimes \overline{\mathbb{Z}}$ be a lifting of $\Lambda(\chi)$, so that

$$\Lambda(\chi) = (S \otimes 1)(\widetilde{\Lambda(\chi)}) \quad .$$

Then

$$\begin{aligned}
(\psi_f \otimes 1)(\Lambda(\chi)) &= \frac{1}{\Omega_f^{\operatorname{sgn}(E)}} \cdot \Lambda(\varphi_f^{\operatorname{sgn}(E)}, \chi) \\
&= \frac{1}{\Omega_f}\left(\frac{(1 + \operatorname{sgn}(E)\, \iota_*)}{2}\, \Lambda(\chi) \cap \varphi_f\right) \\
&= \frac{1}{\Omega_f}\left(\Lambda(\chi) \cap \varphi_f\right) \\
&= \Lambda_f(\chi)
\end{aligned}$$

by Proposition 1.6.3. In particular $\Lambda_f(\chi)$ is $\mathfrak{P}$-integral.

Since $\Omega_{E_\sigma}^{-1} R(E_\sigma) \subseteq \mathfrak{P} \cdot \overline{\mathbb{Z}}_{(\mathfrak{P})}$, Corollary 3.1.5 shows

$$\left(\psi_{E_\sigma} \otimes 1\right) (\widetilde{\Lambda(\chi)}) \equiv \Lambda_{E_\sigma}(\chi) \quad (\mathrm{mod}\ \mathfrak{P}) \quad .$$

Hence $\Lambda_{E_\sigma}(\chi)$ is $\mathfrak{P}$-integral proving the first assertion.

To prove the second assertion:

$$\theta\left(\Lambda_{E_\sigma}(\chi)\right) = \theta \circ \left(\psi_{E_\sigma} \otimes 1\right) (\widetilde{\Lambda(\chi)})$$

$$= \theta \circ (\psi_f \otimes 1)(\Lambda(\chi))$$

$$= \theta(\Lambda_f(\chi)) \quad .$$

□

§4.3. Two Examples: $X_1(13)$, $X_0(7,7)$:

In this section we give two examples of cusp forms f and congruences satisfied by the algebraic parts of their special values $L(f,\chi,1)$.

(1) $X_1(13)$: The genus is 2 and $\mathbb{T} \cong \mathcal{O} = \mathbb{Z}[\rho]$, $\rho = e^{2\pi i/6}$. Let $\varepsilon : \mathbb{Z} \to \mathcal{O}$ be the primitive character of conductor 13 satisfying $\varepsilon(2) = \rho$ and let f be the unique normalized cusp form for $\Gamma_1(13)$ with character ε. Then $A_f = \mathrm{Pic}^0(X_1(13))$.

Let $\pi = 3 + 2\rho \in \mathcal{O}$. Then π is a prime of $\mathcal{O}$ and has norm 19. Let $\mathfrak{P}$ be a prime of $\overline{\mathbb{Q}}$ extending (π).

PROPOSITION 4.3.1: There is a period $\Omega_f^- \in \mathbb{C}^*$ such that for every odd primitive Dirichlet character χ which is nonexceptional at $\mathfrak{P}$ and has conductor prime to 13, the following two properties hold:

1) $$\Lambda_f(\chi) \overset{\text{dfn}}{=} \frac{\tau(\bar{\chi})\, L(f,\chi,1)}{2\pi i\, \Omega_f^-} \in \mathcal{O}[\chi] ,$$

and

2) $$\Lambda_f(\chi) \equiv -\chi(13)\, \varepsilon(m)\, B_1(\chi)\, B_1(\bar{\varepsilon}\bar{\chi}) \pmod{\mathfrak{P}} .$$

Proof: Let ε_0 be the trivial character of conductor 1 and $E = E(\varepsilon, \varepsilon_0)$. We have $B_1(\bar{\varepsilon}) = B_1(\varepsilon_0) = 0$, and $B_2(\bar{\varepsilon}) = 4 \cdot (3+2\rho)/(1+3\rho)$. By 3.4.4 c) we therefore have

$$R(E) = 2 \cdot (3+2\rho)(1+3\rho)^{-1} \cdot \mathcal{O} .$$

In Theorem 3.6.1 we find

$$\mathfrak{a} = \mathcal{O}[\tfrac{1}{2}] \cdot \underline{\text{Periods}}(E) = \mathfrak{b} = (1+3\rho)^{-1} \cdot \mathcal{O}[\tfrac{1}{2}] \quad .$$

Hence

$$A(E) \otimes \mathbb{Z}[\tfrac{1}{2}] \cong \mathcal{O}/\pi\mathcal{O} \quad .$$

In particular $B = C_E \cap A_f[\pi] \neq 0$. Since $\mathcal{O}$ is a Dedekind domain, the first condition of (4.2.2) is satisfied. We can apply Theorem 4.2.3.

Let $\Omega_E \in \underline{\text{Periods}}(E)$ such that $\Omega_E \equiv 1 \pmod{\pi}$. The the image, $\underline{a}$, of Ω_E in $A = B^{\wedge}$ generates A. Let $\Omega_f^- \in \underline{\text{Periods}}(f)^-$ be in the preimage of $\underline{a}$. Let $\sigma \in \mathrm{Gal}(\overline{\mathbb{Q}}/\mathbb{Q})$ be the element guaranteed by Theorem 4.2.3. Since the Nebentypus character of E_σ is congruent to that of f we must have $E_\sigma = E$.

The congruence (2) follows from Theorem 4.2.3 and Proposition 3.4.2 (c). Theorem 4.2.3 guarantees that $\Lambda_f(\chi)$ is $\mathfrak{P}$-integral. But $\underline{\text{Periods}}(f)^-$ is finitely generated so we may modify Ω_f^- by a rational integer factor $\equiv 1 \pmod{\mathfrak{P}}$ so that $\Lambda_f(\chi)$ is actually integral.

(2) $\underline{X_0(7,7)}$: The genus is 1, $\mathbb{T} = \mathbb{Z}$. Let f be the unique normalized cusp form for $\Gamma_0(7,7)$. Let $\pi = 3-\rho \in \mathcal{O} = \mathbb{Z}[\rho]$, $\rho = e^{2\pi i/6}$. Then π is prime in $\mathcal{O}$ and has norm 7. Let $\varepsilon : \mathbb{Z} \to \mathcal{O}$ be the unique primitive Dirichlet character of conductor 7 satisfying $\varepsilon(3) = \rho$. Let $\mathfrak{P}$ be a prime of $\overline{\mathbb{Q}}$ extending (π).

PROPOSITION 4.3.2: There is a period $\Omega_f^+ \in \mathbb{C}^*$ such that for every even primitive Dirichlet character χ of conductor m prime to 7, the following two properties hold:

$$(1) \quad \Lambda_f(\chi) \overset{\text{dfn}}{=} \frac{\tau(\bar{\chi})\, L(f, \chi, 1)}{2\pi i\, \Omega_f^+} \in \mathbf{Z}[\chi] \ ,$$

and $\quad (2) \quad \Lambda_f(\chi) \equiv -\bar{\varepsilon}(m) \cdot \mathbf{B}_1(\varepsilon\chi) \cdot \mathbf{B}_1(\varepsilon\bar{\chi}) \pmod{\mathfrak{P}}$.

<u>Proof</u>: Let $E = E(\bar{\varepsilon}, \varepsilon)$. We can easily compute $\mathbf{B}_1(\varepsilon) = -2\rho/(2+\rho)$, $\mathbf{B}_2(\bar{\varepsilon}, \varepsilon) = 4 \cdot (3-\rho)/(2+\rho)$. Theorem 3.6.1 shows $\mathfrak{O}[\frac{1}{2}] \cdot$ <u>Periods</u> $(E) = (2+\rho)^{-1} \cdot \mathfrak{O}[\frac{1}{2}]$. Proposition 3.4.4 (c) shows $\mathfrak{O}[\frac{1}{2}] \cdot R(E) = \pi \cdot (2+\rho)^{-1} \cdot \mathfrak{O}[\frac{1}{2}]$. Hence

$$A(E) \otimes \mathbf{Z}[\tfrac{1}{2}] \cong \mathfrak{O}/\pi\mathfrak{O} \ .$$

So $B = C_E \cap A_f[7] \neq 0$. The first condition of (4.2.2) is satisfied since A_f is an elliptic curve.

Let $\Omega_E \in$ <u>Periods</u> (E) such that $\Omega_E \equiv 1 \pmod{\pi}$. The image, $\underline{a}$, of Ω_E in $A = B^\wedge$ generates A. Let $\Omega_f^+ \in$ <u>Periods</u> $(f)^+$ be in the preimage of $\underline{a}$. Let $\sigma \in \mathrm{Gal}(\bar{\mathbb{Q}}/\mathbb{Q})$ be as in Theorem 4.2.3.

B. Gross [9] has shown that the elliptic curve $X_0(7, 7)$ (in the guise $X_0(49)$) is the $\mathbb{Q}$-curve $A(7)$. In particular it follows that $h_f(T_\ell) = 0$ if ℓ is a prime other than 7 which is a quadratic nonresidue modulo 7. Then we must have $(\bar{\rho} + 3\rho)^\sigma = h_{E_\sigma}(T_3) \equiv 0 \pmod{\mathfrak{P}}$. So the restriction of σ to $\mathbb{Q}[\rho]$ is trivial and $E_{\sigma} = E$.

The proposition now follows easily from Theorem 4.2.3. $\square$

Chapter 5. P-adic L-functions and Congruences

Mazur and Swinnerton-Dyer [30] have shown how to pass from a T_p-eigenfunction on $(\mathbb{Q}/\mathbb{Z})_{p,\Delta}$ to a p-adic distribution on $\mathbb{Z}^*_{p,\Delta}$. Applying this procedure to the universal modular symbol on a modular curve, X, Mazur [26] defines a p-adic L-function $L_p(X_0(N), \chi, s)$ associated to an Eisenstein prime P for $\Gamma_0(N)$, N prime. He then shows that if χ and $\bar{\chi}$ are nonexceptional, then the image of this p-adic L-function under the Dedekind-Rademacher homomorphism is the reduction modulo $(N-1)$ of

$$(1 - \bar{\chi}(N) \cdot N^{1-s}) \cdot L_p(\chi\omega, 1-s) \cdot L_p(\bar{\chi}\omega, s-1) ,$$

where $L_p(\ )$ is the Kubota-Leopoldt p-adic L-function and ω is the Teichmüller character.

In this chapter we show how this result generalizes to the groups Γ of type (N_1, N_2). For an eigenfunction $E \in \mathcal{E}_2(\Gamma)$ and an Eisenstein prime $P \subseteq \mathbb{T}$ of residual characteristic $p \nmid 2N_1N_2$ we will construct a p-adic L-function $L_p(E, \chi, s)$. This generalizes the Mazur Swinnerton-Dyer construction of $L_p(f, \chi, s)$ associated to a parabolic eigenform $f \in S_2(\Gamma)$. If $E = E(\epsilon_{1,\infty}, \epsilon_{2,\infty})$ then $L_p(E, \chi, s)$ is a product of two Leopoldt-Kubota p-adic L-functions. In case $\epsilon_{1,\infty}, \epsilon_{2,\infty}$ are primitive and $\mathcal{P} \subseteq \mathcal{O}(f)$ satisfies (4.2.2) we prove a congruence between $L_p(f, \chi, s)$

and $L_p(E, \chi, s)$.

In §5.1 we give a basic discussion of p-adic distributions and p-adic L-functions as developed in [41] and [10].

In §5.2 we give two methods of constructing p-adic distributions from special functions $g : (\mathbb{Q}/\mathbb{Z})_{p,\Delta} \to M$. The second of these is due to Mazur and Swinnerton-Dyer [30]. We give formulas for integration of these distributions against characters of finite order in Theorems 5.2.1 and 5.2.2.

The universal p-adic L-function $L_p(X(\Gamma), \chi, s)$ is defined in §5.3 and used to define the p-adic L-function $L_p(f, \chi, s)$ associated to an eigenform f.

The p-adic distribution $\mu_E^{(\Delta)}$ associated to an eigenfunction $E \in \mathcal{E}_2(\Gamma)$ is defined in §5.4. This distribution is not a measure but is "smoothed" to a measure in Proposition 5.4.1. The resulting measure is used to define the p-adic L-function $L_p(E, \chi, s)$.

In §5.5 we describe the modular symbol, f_E, associated to the Eisenstein series $E = E(\epsilon_{1,\infty}, \epsilon_{2,\infty})$ where $\epsilon_{1,\infty}, \epsilon_{2,\infty}$ are primitive. Lemma 5.5.1 is important for the proof of the main theorem of §5.6.

In the last section we prove a "congruence" between the universal p-adic L-function $L_p(X(\Gamma), \chi, s)$ and $L_p(E, \chi, s)$ (Corollary 5.6.6). In case conditions (4.2.2) are satisfied this gives us the desired congruence between $L_p(f, \chi, s)$ and $L_p(E, \chi, s)$ (Corollary 5.6.8).

§5.1. Distributions, Measures and p-adic L-functions:

Throughout this chapter $\mathbb{C}_p$ denotes a fixed completion of $\overline{\mathbb{Q}}_p$ and $\mathbb{D}_p \subseteq \mathbb{C}_p$ is its ring of integers. Let $\mathfrak{O} \subseteq \mathbb{D}_p$ be the ring of integers in a finite extension of $\mathbb{Q}_p$ and M be a topological $\mathfrak{O}$-module.

Let G be an abelian profinite group. The Iwasawa algebra of G over $\mathfrak{O}$ is ([41] and [10])

$$\Lambda_G = \Lambda_G(\mathfrak{O}) \overset{\text{dfn}}{=} \mathfrak{O}[[G]] = \varprojlim_{\substack{H \subseteq G \\ \text{open subgroup}}} \mathfrak{O}[G/H] \quad .$$

An element $\lambda \in \Lambda_G$ is called an $\mathfrak{O}$-valued distribution on G. Let $M[G/H] = M \otimes_{\mathfrak{O}} \mathfrak{O}[G/H]$ and define the Λ_G-module of M-valued distributions on G to be

$$M[[G]] \overset{\text{dfn}}{=} \varprojlim_{H} M[G/H] \quad .$$

Let Step$(G;\mathfrak{O})$ be the module of locally constant $\mathfrak{O}$-valued functions on G. If $f \in$ Step$(G;\mathfrak{O})$ is constant on cosets of an open subgroup $H \subseteq G$ and if $\mu \in M[[G]]$ has image $\sum_{s \in G/H} \mu(s) \cdot s$ in $M[G/H]$ then define the integral

$$\langle f, \mu \rangle = \int_G f\,d\mu \overset{\text{dfn}}{=} \sum_{s \in G/H} f(s) \cdot \mu(s) \in M \quad .$$

This is independent of the choice of H.

We call $\mu \in M[[G]]$ a measure on G if the image of integration $\langle -, \mu \rangle : \underline{\text{Step}}(G; \mathcal{O}) \to M$ is finitely generated over $\mathcal{O}$. Let

$$\Lambda_G(M) = \{M\text{-valued measures on } G\} \quad .$$

This is a Λ_G-submodule of $M[[G]]$.

Integration with respect to $\mu \in \Lambda_G(M)$ is continuous for the topology of uniform convergence on $\underline{\text{Step}}(G; \mathcal{O})$ and hence extends uniquely to a continuous $\mathcal{O}$-linear map on the module of continuous $\mathcal{O}$-valued functions on G:

$$\begin{array}{ccc} \underline{\text{Cont}}(G; \mathcal{O}) & \longrightarrow & M \\ f & \longmapsto & \langle f, \mu \rangle = \int_G f \, d\mu \end{array} \quad .$$

A homomorphism of $\mathcal{O}$-modules $\varphi : M_1 \to M_2$ induces a Λ_G-homomorphism $\varphi_* : M_1[[G]] \to M_2[[G]]$ satisfying the integration formula

$$\langle f, \varphi_* \mu \rangle = \varphi(\langle f, \mu \rangle) \quad , \quad \text{for } f \in \underline{\text{Step}}(G; \mathcal{O}) \quad .$$

A homomorphism $\phi : G \to H$ of profinite groups induces a continuous Λ_G-homomorphism $\phi^* : M[[G]] \to M[[H]]$ satisfying for $f \in \underline{\text{Step}}(H; \mathcal{O})$ and $\mu \in M[[G]]$

$$\langle f, \phi^* \mu \rangle_H = \langle f \circ \phi, \mu \rangle \quad .$$

If $i : \mathcal{O} \hookrightarrow \mathcal{O}'$ is a finite extension of discrete valuation rings in $\mathbf{D}_p$ then we obtain a base change homomorphism $i_* : M[[G]] \to (M \otimes_{\mathcal{O}} \mathcal{O}')[[G]]$

which satisfies for $f \in \underline{\mathrm{Step}}(G;\mathfrak{O})$, $\mu \in M[[G]]$

$$\langle i \circ f, i_*(\mu)\rangle = (1 \otimes i)(\langle f, \mu\rangle) \quad .$$

For a character $\chi : G \to \mathbb{D}_p^*$ of finite order, let $\mathfrak{O}[\chi] \subseteq \mathbb{D}_p$ be the discrete valuation ring generated over $\mathfrak{O}$ by the values of χ. Let $i = i(\chi) : \mathfrak{O} \hookrightarrow \mathfrak{O}[\chi]$ and $M[\chi] = M \otimes_{\mathfrak{O}} \mathfrak{O}[\chi]$. Define

$$R_\chi : M[[G]] \longrightarrow M[\chi][[G]]$$

by the integration formula $\langle f \cdot \chi, i_*\mu\rangle = \langle f, R_\chi(\mu)\rangle$ for $f \in \underline{\mathrm{Step}}(G;\mathfrak{O}[\chi])$.

The homomorphisms, φ_*, ϕ^*, i_*, R_χ of the last four paragraphs are measure preserving.

The "dual distribution" to $\mu \in M[[G]]$ is the distribution $\hat{\mu} = \phi^*(\mu)$ where $\phi : G \to G$ is defined by $\varphi(g) = g^{-1}$.

In the rest of this section we restrict our attention to the special case of interest to us.

Let $\Delta > 0$ be an integer such that $p \nmid \Delta$. Set

$$G = G^{(\Delta)} \overset{\mathrm{dfn}}{=} \mathbb{Z}^*_{p,\Delta} = \varprojlim_n (\mathbb{Z}/p^n \Delta \mathbb{Z})^* \quad .$$

Let $q = p$ if $p \neq 2$ and $q = 4$ if $p = 2$, and let $U \subseteq \mathbb{Z}_p^*$ be the subgroup of elements congruent to $1 \bmod q$. The group G splits canonically as

$$G \cong (\mathbb{Z}/q\Delta)^* \times U \quad .$$

Let

$$\phi_U : G \longrightarrow U$$

be projection to the second factor.

Let $\chi : G \to \mathbb{C}_p^*$ be a continuous character. If χ factors through projection to $(\mathbb{Z}/q\Delta)^*$ we call χ tame. If it factors through ϕ_U we call χ wild. An arbitrary character χ can be uniquely expressed as a product of a tame and a wild character.

By composing ϕ_U with the natural embedding $U \subseteq \mathbb{C}_p^*$ we obtain a wild character

$$\langle \quad \rangle : G \longrightarrow \mathbb{C}_p^* \quad .$$

For $s \in \mathbb{Z}_p$, $x \in G$ let $\langle x \rangle^s = \exp(s \log \langle x \rangle)$.

Let $\chi : G \to \mathbb{C}_p^*$ be a character of finite order and define

$$\phi_\chi : M[[G]] \longrightarrow M[\chi][[U]]$$

to be the composition $\phi_\chi = \phi_U^* \circ R_\chi$.

We define the p-adic L-function attached to a measure $\mu \in \Lambda_G(M)$ and a character $\chi : G \to \mathbb{C}_p^*$ of finite order to be

$$L_p(\mu, \chi, s) \overset{\text{dfn}}{=} \langle \langle \quad \rangle^{s-1}, \phi_\chi(\mu) \rangle$$

which we view as an $M[\chi]$-valued function of the variable $s \in \mathbb{Z}_p$.

Observe the special value

$$L_p(\mu, \chi, 1) = \langle \chi, i(\chi)_* \mu \rangle \quad .$$

PROPOSITION 5.1.1: Let M be a free $\mathcal{O}$-module of finite rank, and $\mu \in \Lambda_U(M)$ be a measure on U. If

$$L_p(\mu, \chi, 1) = 0$$

for all nontrivial characters $\chi : U \to \mathbb{C}_p^*$ of finite order then $\mu = 0$.

Proof: It suffices to show $\langle f, \mu \rangle = 0$ for all $f \in \underline{\text{Step}}(U; \mathcal{O})$. We do this first when $f = 1$ is the trivial character.

Let $U_n \subseteq U$ be the group of units congruent to 1 modulo qp^n. Then $U/U_n \cong (\mathbb{Z}/p^n\mathbb{Z})$. Let X_n be the group of $\mathbb{C}_p^*$-valued characters on U which are trivial on U_n. Then

$$\sum_{\chi \in X_n} \chi = p^n \cdot \delta_n$$

where δ_n is the characteristic function of U_n. Let $i_n : \mathcal{O} \hookrightarrow \mathcal{O}[\zeta_n]$ where $\zeta_n \in \mathbb{C}_p^*$ is a primitive p^n-th root of 1. Then $(1 \otimes i_n) : M \hookrightarrow M_n \overset{\text{dfn}}{=} M \otimes \mathcal{O}[\zeta_n]$ and

$$\begin{aligned}(1 \otimes i_n)(\langle 1, \mu \rangle) &= \langle 1, i_{n*}(\mu) \rangle = p^n \langle \delta_n, i_{n*}(\mu) \rangle \\ &\equiv 0 \pmod{p^n \cdot M_n} \quad .\end{aligned}$$

Hence $\langle 1, \mu \rangle \equiv 0 \pmod{p^n M}$. Since this is true for all n we have $\langle 1, \mu \rangle = 0$.

If $f \in \underline{\text{Step}}(U; \mathcal{O})$ is constant on cosets of U_n, then

$p^n \cdot f = \sum_{\chi \in X_n} a_\chi \cdot \chi$ for an appropriate choice of $a_\chi \in \mathfrak{O}[\zeta_n]$. So

$$(1 \otimes i_n)(p^n \cdot \langle f, \mu \rangle) = p^n \cdot \sum_{\chi} a_\chi \cdot \langle \chi, i_{n*}(\mu) \rangle = 0$$

shows $\langle f, \mu \rangle = 0$. $\square$

§5.2. Construction of Distributions:

For $n \geq 0$ let $G_n = (\mathbb{Z}/p^n \Delta \mathbb{Z})^*$ and $\phi_n : G \to G_n$ be the natural projection. For each $r \in \mathbb{Z}$ with $(r, p\Delta) = 1$ let $\sigma(r) \in G$ be the corresponding element of $\mathbb{Z}^*_{p,\Delta}$ and let $\sigma_n(r) = \phi_n(\sigma(r)) \in G_n$.

A distribution $\mu \in M[[G]]$ determines a sequence of functions $\mu_n : (\mathbb{Z}/p^n \Delta \mathbb{Z})^* \to M$, $n \geq 0$, by

$$\phi^*_n(\mu) = \sum_{a \in (\mathbb{Z}/p^n \Delta \mathbb{Z})^*} \mu_n(a) \cdot \sigma_n(a) \in M[G_n] \quad .$$

These satisfy the distribution relations

$$\mu_n(a) = \sum_{\substack{b \in (\mathbb{Z}/p^{n+1} \Delta \mathbb{Z})^* \\ b \equiv a \pmod{p^n \Delta}}} \mu_{n+1}(b)$$

for $a \in (\mathbb{Z}/p^n \Delta \mathbb{Z})^*$, and $n \geq 0$. Conversely such a sequence determines a distribution.

Let $(\mathbb{Q}/\mathbb{Z})_{p,\Delta}$ be the set of elements of $\mathbb{Q}/\mathbb{Z}$ which may be represented by a rational number of the form $\frac{a}{p^n \Delta}$ with $(a, p\Delta) = 1$. Let $\mathfrak{F} = \mathfrak{F}((\mathbb{Q}/\mathbb{Z})_{p,\Delta}; M)$ be the $\mathfrak{O}$-module of functions $g : (\mathbb{Q}/\mathbb{Z})_{p,\Delta} \to M$. We say g is bounded if its image is contained in a finitely generated $\mathfrak{O}$-submodule of M. The group $G = \mathbb{Z}^*_{p,\Delta}$ acts on $(\mathbb{Q}/\mathbb{Z})_{p,\Delta}$ by $\sigma(r) \cdot \frac{a}{p^n \Delta} = \frac{ar}{p^n \Delta}$ for $(r, p\Delta) = 1$. We give $\mathfrak{F}$ the structure of Λ_G-module by defining

$$(\sigma(r) \cdot g)(x) = g(\sigma(r)^{-1} \cdot x)$$

The bounded functions form a Λ_G-submodule of $\mathfrak{F}$.

We associate to a distribution $\mu \in M[[G]]$ the function $g_\mu \in \mathfrak{F}$ by

$$g_\mu\left(\frac{a}{p^n \Delta}\right) = \mu_n(a) \quad .$$

The map

$$\begin{aligned} M[[G]] &\longrightarrow \mathfrak{F} \\ \mu &\longmapsto g_\mu \end{aligned}$$

is a Λ_G-homomorphism. The distribution μ is a measure if and only if g_μ is bounded.

We will identify a primitive Dirichlet character $\chi : \mathbb{Z} \to \mathbb{C}_p$ of conductor $p^n \Delta$ with the associated character $\chi : G \to \mathbb{C}_p^*$. If $g \in \mathfrak{F}$ let

$$\mathcal{L}(g, \chi) \overset{\text{dfn}}{=} \sum_{a \in (\mathbb{Z}/p^n \Delta \mathbb{Z})^*} g\left(\frac{a}{p^n \Delta}\right) \otimes \chi(a) \in M[\chi] \quad .$$

For $\mu \in M[[G]]$ we have

$$\langle \chi, \mu \rangle = \mathcal{L}(g_\mu, \chi) \quad .$$

We now examine two circumstances (I and II below) under which we can construct distributions using functions in $\mathfrak{F}$. The ideas here are due to Mazur [30], but in II we replace his limiting process with a convenient explicit formula.

(I) Define the operator S on $\mathfrak{F}$ by

$$(g|S)(x) = \sum_{\substack{y \in (\mathbb{Q}/\mathbb{Z})_{p,\Delta} \\ py \equiv x \pmod 1}} g(y) \quad .$$

This operator commutes with the Λ_G-action. Let $\rho \in \mathrm{End}_{\mathfrak{G}}(M)$ and $\mathfrak{F}^{[\rho S = 1]}$ be the set of $g \in \mathfrak{F}$ such that $\rho \cdot g|S = g$.

PROPOSITION 5.2.1: Let $g \in \mathfrak{F}^{[\rho S = 1]}$ and define $\mu_n : (\mathbb{Z}/p^n \Delta \mathbb{Z})^* \to M$, $n \geq 0$ by

$$\mu_n(a) = \begin{cases} \rho^n \cdot g\left(\dfrac{a}{p^n \Delta}\right) & \text{if} \quad n > 0 \ , \\[2ex] g\left(\dfrac{a}{\Delta}\right) - \rho \cdot g\left(\dfrac{p' a}{\Delta}\right) & \text{if} \quad n = 0 \ , \end{cases}$$

where p' is an integer satisfying $pp' \equiv 1 \pmod{\Delta}$. Then

(a) $\{\mu_n\}_{n=0}^{\infty}$ satisfies the distribution laws and therefore defines a distribution $\mu \in M[[G]]$.

(b) The map $g \mapsto \mu$ defines a Λ_G-homomorphism

$$\mathfrak{F}^{[\rho S = 1]} \longrightarrow M[[G]] \quad .$$

(c) If $\chi : \mathbb{Z} \to \mathbb{C}_p$ is a primitive Dirichlet character of conductor $p^n\Delta$, $n \geq 0$, then

$$\langle \chi, \mu \rangle = \rho^n (1 - \chi(p)\rho)\, \mathcal{L}(g, \chi) \in M[\chi] \quad .$$

Proof: (b) and (c) are immediate from the explicit formula for μ. We verify only (a). Let $n \geq 0$ and $a \in (\mathbb{Z}/p^n \Delta \mathbb{Z})^*$. Then

$$\sum_{b \in (\mathbb{Z}/p^{n+1}\Delta\mathbb{Z})^*} \mu_{n+1}(b) = \rho^{n+1} \cdot \sum_b g\left(\frac{b}{p^{n+1}\Delta}\right) =$$

$$= \rho^{n+1} \cdot \sum_{\substack{k=0 \\ p \nmid (a + p^n \Delta k)}}^{p-1} g\left(\frac{a + p^n \Delta k}{p^{n+1}\Delta}\right)$$

$$= \begin{cases} \rho^{n+1} \cdot (g|S)\left(\dfrac{a}{p^n \Delta}\right) & \text{if} \quad n > 0 \,, \\ \rho \cdot (g|S)\left(\dfrac{a}{\Delta}\right) - \rho \cdot g\left(\dfrac{p'a}{\Delta}\right) & \text{if} \quad n = 0 \quad . \end{cases}$$

Since $\rho \cdot g|S = g$ this is $\mu_n(a)$. □

(II) Let $\varepsilon \in \operatorname{End}_{\mathfrak{G}}(M)$ and define the operator π and the "Hecke operator" $T_{p,\varepsilon}$ on $\mathfrak{F}$ by the formulae

$$(g|\pi)(x) = g(px)$$

$$g|T_{p,\varepsilon} = g|S + \varepsilon \cdot g|\pi \quad .$$

These operators commute with the action of Λ_G. Let $t \in \operatorname{End}_{\mathfrak{G}}(M)$ and suppose $g \in \mathfrak{J}^{[T_{p,\varepsilon} = t]}$. Then

$$t \cdot g|S = g|(S + \varepsilon\pi)S = g|(S^2 + p\varepsilon)$$

since $\pi S = p$. We write this in the suggestive form

$$g|(S^2 - tS + p\varepsilon) = 0 \quad .$$

Suppose $\rho \in \mathrm{End}_{\mathfrak{G}}(M)$ satisfies

$$p\rho^2\varepsilon - \rho t + 1 = 0 \quad .$$

Then $0 = g|\rho^2(S^2 - tS + p\varepsilon) = g|(\rho S - p\rho^2\varepsilon)(\rho S - 1)$. So we arrive at a distribution by applying to g the following composition of Λ_G-homomorphisms

$$\mathfrak{J}^{[T_{p,\varepsilon} = t]} \xrightarrow{(\rho S - p\rho^2\varepsilon)} \mathfrak{J}^{[\rho S = 1]} \longrightarrow M[[G]]$$

$$g \longmapsto \mu$$

PROPOSITION 5.2.2: Let $\rho, t, \varepsilon \in \mathrm{End}_{\mathfrak{G}}(M)$ be related by the equation

$$p\rho^2\varepsilon - \rho t + 1 = 0 \quad .$$

Let $g \in \mathfrak{J}^{[T_{p,\varepsilon} = t]}$ and $\mu \in M[[G]]$ be the distribution associated to g by the preceding discussion. Let $\chi : \mathbb{Z} \to \mathbb{C}_p$ be a primitive Dirichlet character of conductor $p^n\Delta$, $n \geq 0$. Then

$$\langle \chi, \mu \rangle = \rho^n(1 - \chi(p)\rho)(1 - \bar{\chi}(p)\rho\varepsilon)\,\mathfrak{L}(g, \chi) \quad .$$

<u>Proof:</u> By 5.2.1

$$\langle \chi, \mu \rangle = \rho^n(1 - \chi(p)\rho) \cdot \mathfrak{L}(h, \chi)$$

where $h = \rho g|S - p\rho^2 \varepsilon \cdot g$. But $g|S = t \cdot g - \varepsilon \cdot g|\pi$ so $h = (\rho t - p\rho^2 \varepsilon) g - \rho\varepsilon\, g|\pi = g - \rho\varepsilon \cdot g|\pi$. From this formula we see $\mathcal{L}(h, \chi) = (1 - \bar{\chi}(p)\rho\varepsilon)\, \mathcal{L}(g, \chi)$. □

We illustrate (I) with a standard example. Let $\mathbf{B}_1 : (\mathbb{Q}/\mathbb{Z})_{p,\Delta} \longrightarrow \mathbb{Q} \hookrightarrow \mathbb{Q}_p$ be the Bernoulli function. Then $\mathbf{B}_1 \in \mathfrak{F}((\mathbb{Q}/\mathbb{Z})_{p,\Delta}; \mathbb{Q}_p)^{[S=1]}$. Let $\lambda_{\mathbf{B}}^{(\Delta)} \in \mathbb{Q}_p[[G^{(\Delta)}]]$ be the distribution associated to $\mathbf{B}_1$ by (I). If $\chi : \mathbb{Z} \to \mathbb{C}_p$ has conductor $p^n\Delta$ then by 5.2.1

$$\langle \chi, \lambda_{\mathbf{B}}^{(\Delta)} \rangle = (1 - \chi(p)) \cdot \mathbf{B}_1(\chi) \in \mathbb{Q}_p[\chi] \quad .$$

The distribution $\lambda_{\mathbf{B}}$ is not a measure but can easily be "smoothed." Let $r > 1$ be prime to $p\Delta$ and let

$$Sm_r \lambda_{\mathbf{B}} = (1 - r\sigma(r)) \cdot \lambda_{\mathbf{B}} \quad .$$

We claim $Sm_r \lambda_{\mathbf{B}} \in \Lambda_G(\mathbb{Z}_p)$. It suffices to show $(1 - r\sigma(r)) \cdot \mathbf{B}_1$ is $\mathbb{Z}_p$-valued. Let $n \geq 0$ and $r' \in \mathbb{Z}$ be such that $rr' \equiv 1 \pmod{p^n\Delta}$. Then

$$\begin{aligned}(1 - r\sigma(r))\, \mathbf{B}_1\left(\frac{a}{p^n\Delta}\right) &= \mathbf{B}_1\left(\frac{a}{p^n\Delta}\right) - r \cdot \mathbf{B}_1\left(\frac{ar'}{p^n\Delta}\right) \\ &\equiv 0 \pmod{\mathbb{Z}_p} \quad .\end{aligned}$$

Let $\chi : \mathbb{Z} \to \mathbb{C}_p$ be primitive of conductor $p^n\Delta$, $n \geq 0$. We will write $\lambda_{\mathbf{B}}(\chi)$ for $\phi_\chi(\lambda_{\mathbf{B}}) \in \mathbb{Q}_p[\chi][[U]]$.

PROPOSITION 5.2.3: Suppose χ is nonexceptional at p in the sense of 3.1.3. Then

$$\lambda_{\mathbf{B}}(\chi) \in \Lambda_U(\mathbb{Z}_p[\chi]) \quad .$$

<u>Proof:</u> We have $(1 - r\chi(r)\langle r\rangle)\lambda_{\mathbf{B}}(\chi) = \phi_\chi(\mathrm{Sm}_r\,\lambda_{\mathbf{B}}) \in \Lambda_U(\mathbb{Z}_p[\chi])$. Since χ is nonexceptional we can choose r such that $1 - r\chi(r)$ is prime to p. For this choice of r, $(1 - r\chi(r)\langle r\rangle)$ is invertible in $\Lambda_U(\mathbb{Z}_p[\chi])$. □

We define the p-adic L-function associated to the pair $\mathbf{B}_1, \chi$ by

$$L_p(\mathbf{B}_1, \chi, s) \overset{\mathrm{dfn}}{=} \frac{1}{(1 - r\chi(r)\langle r\rangle^{s-1})} \cdot \langle\langle \ \ \rangle^{s-1}, \phi_\chi(\mathrm{Sm}_r\,\lambda_{\mathbf{B}}^{(\Delta)})\rangle \quad .$$

This is independent of the choice of r satisfying $(r, p\Delta) = 1$ and $r\chi(r)\langle r\rangle^{s-1} \neq 1$.

This p-adic L-function is related to the Kubota-Leopoldt p-adic L-function by

$$L_p(\mathbf{B}_1, \chi, s) = L_p(\chi\omega, 1 - s)$$

where $\omega : (\mathbb{Z}/p\mathbb{Z})^* \to \mathbb{Z}_p^*$ is the Teichmüller character.

§5.3: Universal measures and measures associated to cusp forms:

Let $N_1, N_2 > 0$, $N = \mathrm{lcm}(N_1, N_2)$, Γ be of type (N_1, N_2) and $X = X(\Gamma)$. Mazur [26] defines the universal modular symbol attached to Γ to be the function

$$\underline{\mathrm{Univ}} : \mathbb{P}^1(\mathbb{Q}) \longrightarrow H_1(X, \underline{\mathrm{cusps}}; \mathbb{Z})$$
$$x \longmapsto \{i\infty, x\}_\Gamma \quad .$$

Then $\underline{\mathrm{Univ}}(\gamma x) - \underline{\mathrm{Univ}}(x) = [\gamma] \in H_1(X; \mathbb{Z})$ for $\gamma \in \Gamma$.

More generally a function $g : \mathbb{P}^1(\mathbb{Q}) \to M$ into an abelian group M is called a modular symbol for Γ if

(i) $g(i\infty) = 0$,

(ii) For $x, y \in \mathbb{P}^1(\mathbb{Q})$, $\gamma \in \Gamma$

$$g(\gamma x) - g(x) = g(\gamma y) - g(y) \quad ..$$

Every modular symbol factors through the universal modular symbol ([26], II §1).

In the proof of 1.8.1 we showed (in its dual form) that the exact sequence of $\mathfrak{J}$-modules

$$0 \longrightarrow H_1(X; \mathbb{Q}) \longrightarrow H_1(X, \underline{\mathrm{cusps}}; \mathbb{Q}) \longrightarrow \widetilde{H}_0(\mathrm{cusps}; \mathbb{Q}) \longrightarrow 0$$

splits canonically. Let $\eta : H_1(X, \underline{\mathrm{cusps}}; \mathbb{Q}) \to H_1(X; \mathbb{Q})$ be the associated projection, and let $H^\partial \subseteq H_1(X; \mathbb{Q})$ be the image of $H_1(X, \underline{\mathrm{cusps}}; \mathbb{Z})$. Then

H^{∂} is a finitely generated $\mathbb{T}$-module.

Let $P \subseteq \mathbb{T}$ be a fixed ordinary prime of residue characteristic $p \nmid N$. Let $\mathbb{T}_P$ be the completion of $\mathbb{T}$ at P and $H_P^{\partial} = H^{\partial} \otimes_{\mathbb{T}} \mathbb{T}_P$. Let $t_p, \epsilon_p \in \mathbb{T}_P \subseteq \mathrm{End}_{\mathbb{Z}_p}(H_P^{\partial})$ correspond to the Hecke operators T_p, $\langle p \rangle$ respectively.

Define a function $g_P \in \mathfrak{F}(\mathbb{Q}/\mathbb{Z}; H_P^{\partial})$ by

$$g_P(x) = \eta_P \circ \underline{\mathrm{Univ}}(N_2 x) \quad .$$

Then $(g_P | T_{p,\epsilon})(x) = \eta_P \left(\sum_{k=0}^{p-1} \mathrm{Univ}\left(\frac{N_2 x + N_2 k}{p} \right) + \epsilon_p \mathrm{Univ}(N_2 p x) \right) =$

$$= \eta_P \ \pi_\Gamma \left(\sum_{k=0}^{p-1} \begin{pmatrix} 1 & N_2 k \\ 0 & p \end{pmatrix} \cdot \{i\infty, N_2 x\} + \sigma_p \begin{pmatrix} p & 0 \\ 0 & 1 \end{pmatrix} \cdot \{i\infty, N_2 x\} \right) =$$

$= t_p \cdot g_P(x)$. Hence $g_P | T_{p,\epsilon_p} = t_p \cdot g_P$.

Let $\rho_p \in \mathbb{T}_P$ be the unique root of the polynomial $p \cdot \epsilon_p \cdot x^2 - t_p x + 1 \in \mathbb{T}_P[x]$. The existence and uniqueness follows from Hensel's Lemma since t_p is a unit in $\mathbb{T}_P$.

Using II of §5.2 we can construct a measure $\mu_P^{(\Delta)} \in H_P^{\partial}[[G^{(\Delta)}]]$ for each $\Delta > 0$ prime to p. If $\pm$ denotes either $+$ or $-$, then we define the measure $\mu_P^{(\Delta),\pm}$ to be the measure obtained from $\mu_P^{(\Delta)}$ by the projection $H_P^{\partial} \twoheadrightarrow H_P^{\partial,\pm}$.

Let $1 \neq \chi : \mathbb{Z} \to \mathbb{C}_p$ be a primitive Dirichlet character of conductor $p^n \Delta$, $n \geq 0$ with $(p\Delta, N) = 1$. Let $H_P^{\partial,\pm}[\chi] = H_P^{\partial,\pm} \otimes_{\mathbb{Z}_p} \mathbb{Z}_p[\chi]$, and let

$$\mu_P(\chi) \stackrel{\text{dfn}}{=} \phi_\chi\left(\mu_P^{(\Delta),\, \operatorname{sgn}\chi}\right) \in H_P^{\partial,\, \operatorname{sgn}\chi}[\chi][[U]] \quad .$$

<u>Definition 5.3.1</u>: The universal P-adic L-function associated to χ is

$$L_P(X, \chi, s) \stackrel{\text{dfn}}{=} \langle \langle \quad \rangle^{s-1}, \mu_P(\chi) \rangle$$

which we view as an $H_P^{\partial,\operatorname{sgn}\chi}[\chi]$-valued function of the variable $s \in \mathbb{Z}_p$. □

We fix for the rest of the chapter a prime $p \nmid N$ and an isomorphism $\kappa : \mathbb{C} \xrightarrow{\sim} \mathbb{C}_p$. Let $\chi_\infty = \kappa^{-1} \circ \chi : \mathbb{Z} \to \mathbb{C}$ and $\Lambda_P(\chi)$ be the image of the universal special value $\Lambda(\chi_\infty)$ under the natural map $H_1(X; \mathbb{Z}[\chi_\infty]) \to$ $\to H_P^{\operatorname{sgn}(\chi)}[\chi]$.

The next proposition expresses the relationship between the special value at $s = 1$ of the universal P-adic L-function and the universal special value $\Lambda(\chi_\infty)$.

THEOREM 5.3.2:

$$L_P(X, \chi, 1) = \bar{\chi}(N_2) \cdot \rho_p^n (1 - \chi(p)\rho_p)(1 - \bar{\chi}(p)\rho_p \epsilon_p)\Lambda_P(\bar{\chi}) \in H_P^{\operatorname{sgn}(\chi)}[\chi] \quad .$$

<u>Proof</u>: This is a simple application of 5.2.2 and the formula

$$\chi(N_2) \cdot \mathcal{L}(g_P, \chi) = \Lambda_P(\bar{\chi}) . \quad \square$$

Let f be a weight 2 cusp form on Γ, and suppose f is an eigenfunction for the full Hecke algebra $\mathbb{T}$. Let $h = \kappa \circ h_f : \mathbb{T} \to \mathbb{D}_p$ be the corresponding homomorphism and $\varphi = \kappa \circ \varphi_f : H^\partial \to \mathbb{C}_p$ be the $\mathbb{T}$-homomorphism defined by integration with respect to $f(z)\,dz$. Let $P \subseteq \mathbb{T}$ be the pullback of the maximal ideal in $\mathbb{D}_p$, and suppose as before that P is ordinary. Let

$$h_P : \mathbb{T}_P \longrightarrow \mathbb{D}_p$$

$$\varphi_P : H_P^\partial \longrightarrow \mathbb{C}_p$$

be the completions of h and φ respectively.

For each $\Delta > 0$ with $p \nmid \Delta$ define the measure

$$\mu_f^{(\Delta)} \stackrel{dfn}{=} \varphi_{P,*}(\mu_P^{(\Delta)}) \in \Lambda_{G^{(\Delta)}}(\mathbb{C}_p) .$$

If $\chi : \mathbb{Z} \to \mathbb{C}_p$ is a primitive Dirichlet character of conductor $p^n \Delta$, $n \geq 0$, let

$$\mu_f(\chi) = \phi_\chi(\mu_f^{(\Delta)}) \in \Lambda_U(\mathbb{C}_p) ,$$

and define the p-adic L-function

$$L_p(f, \chi, s) \overset{\text{dfn}}{=} \langle \langle \ \rangle^{s-1}, \mu_f(\chi) \rangle$$

which is a $\mathbb{C}_p$-valued function of $s \in \mathbb{Z}_p$.

Let $\rho = h_p(\rho_p)$ and $\epsilon = h_p(\epsilon_p)$. The following theorem is immediate from 5.3.2 and Proposition 1.6.3.

THEOREM 5.3.3: Suppose the conductor of χ is prime to N. Let $\chi_\infty = \kappa^{-1} \circ \chi$. Then

$$\chi(N_2)\, L_p(f, \chi, 1) = \rho^n (1 - \chi(p)\rho)(1 - \bar{\chi}(p)\rho\,\epsilon) \cdot \left(\frac{\tau(\chi_\infty)\, L(f, \bar{\chi}_\infty, 1)}{2\pi i} \right)^{\kappa} .$$

□

§5.4. Measures associated to Eisenstein Series:

Let Γ be of type (N_1, N_2) as in the last section. Let $\varepsilon_1, \varepsilon_2 : \mathbb{Z} \to \mathbb{C}_p$ be not necessarily primitive Dirichlet characters of conductors N_1, N_2 respectively, and set $\varepsilon_{i,\infty} = \kappa^{-1} \circ \varepsilon_i$, $i = 1, 2$. Let $K = \mathbb{Q}[\varepsilon_{1,\infty}, \varepsilon_{2,\infty}]$ and let $E \in \mathcal{E}_2(\Gamma; K)$ be a nonzero weight two Eisenstein series of signature $(\varepsilon_{1,\infty}, \varepsilon_{2,\infty})$ which is an eigenfunction for the full abstract Hecke algebra $\mathfrak{J}$. Finally, let $\varepsilon = \varepsilon_1 \varepsilon_2$ and $\varepsilon_\infty = \varepsilon_{1,\infty} \varepsilon_{2,\infty}$.

Let $K_p = \mathbb{Q}_p[\varepsilon_1, \varepsilon_2]$ and $\mathcal{O} = \mathbb{Z}_p[\varepsilon_1, \varepsilon_2]$ be the ring of integers in K_p.

For $\Delta > 0$ with $(\Delta, pN) = 1$ define $g_E^{(\Delta)} \in \mathfrak{J}((\mathbb{Q}/\mathbb{Z})_{p,\Delta}; K_p)$ by $g_E^{(\Delta)}(x) = \kappa s_E(N_2 x)$, $x \in (\mathbb{Q}/\mathbb{Z})_{p,\Delta}$, where s_E is the Dedekind symbol of 2.5.2. Define

$$\tilde{e}_E : GL_2^+(\mathbb{Q}) \longrightarrow K_p$$

by $\tilde{e}_E(\gamma) = \frac{1}{2} \kappa(e_E(\gamma) + e_E^{\iota}(\gamma))$ for $\gamma \in GL_2^+(\mathbb{Q})$. Then, for $x \in (\mathbb{Q}/\mathbb{Z})_{p,\Delta}$

$$g_E^{(\Delta)}(x) = \tilde{e}_E\begin{pmatrix} 1 & N_2 x \\ 0 & 1 \end{pmatrix} .$$

Let $\chi : \mathbb{Z} \to \mathbb{C}_p$ be a nontrivial primitive Dirichlet character of conductor $m = p^n \Delta$, $n \geq 0$, and set $\chi_\infty = \kappa^{-1} \circ \chi$. Then by Lemma 3.1.1

$$\chi(N_2)\cdot \mathfrak{L}(g_E^{(\Delta)},\chi) = \sum_{a=1}^{m-1} \chi(N_2 a)\,\widetilde{e}_E\begin{pmatrix}1 & N_2 a\\ 0 & m\end{pmatrix}$$

$$= \frac{1}{2}(1-\chi\epsilon_1(-1)) \sum_{a=1}^{m-1} \chi(N_2 a)\, e_E\begin{pmatrix}1 & N_2 a\\ 0 & m\end{pmatrix}^{\kappa}$$

$$= \begin{cases} \left(\dfrac{-\tau(\chi_\infty)\,L(E,\overline{\chi}_\infty,1)}{2\pi i}\right)^{\kappa} & \text{if } \operatorname{sgn}\chi = \operatorname{sgn}(E), \\ 0 & \text{otherwise}. \end{cases}$$

Let $t = \epsilon_1(p) + p\epsilon_2(p) \in \mathfrak{O}^*$. Then $(g_E^{(\Delta)}|T_{p,\epsilon(p)})(x) =$

$$= \sum_{k=0}^{p-1} \widetilde{e}_E\begin{pmatrix}1 & N_2(x+k)\\ 0 & p\end{pmatrix} + \widetilde{e}_{E|\langle p\rangle}\begin{pmatrix}p & N_2 px\\ 0 & 1\end{pmatrix} = \widetilde{e}_{E|T_p}\begin{pmatrix}1 & N_2 x\\ 0 & 1\end{pmatrix} =$$

$$= t\cdot g_E^{(\Delta)}(x).$$

The unique root in $\mathfrak{O}$ of the polynomial $p\epsilon(p)x^2 - tx + 1 \in \mathfrak{O}[x]$ is $\overline{\epsilon}_1(p)$. So using II of §5.2 we obtain a distribution $\mu_E^{(\Delta)} \in K_p[[G^{(\Delta)}]]$, which satisfies the integration formula

$$\langle \chi, \mu_E^{(\Delta)}\rangle = \overline{\epsilon}_1(p^n)(1-\overline{\epsilon}_1\chi(p))(1-\epsilon_2\overline{\chi}(p))\left(\frac{-\tau(\chi_\infty)\,L(E,\overline{\chi}_\infty,1)}{2\pi i}\right)^{\kappa}$$

if $\operatorname{sgn}(\chi) = \operatorname{sgn}(E)$.

Though $\mu_E^{(\Delta)}$ is not a measure it can be smoothed as follows. Let $r_1, r_2 \in \mathbb{Z}$ satisfy $(r_1 r_2, p\Delta) = 1$ and let $\sigma(r_i) \in G^{(\Delta)}$ be the corresponding group elements. Define

$$\mathrm{Sm}_{r_1}^{r_2}(\mu_E^{(\Delta)}) \overset{\mathrm{dfn}}{=} (1 - r_1\,\sigma(r_1))\,(1 - r_2\,\sigma(r_2)^{-1})\,\mu_E^{(\Delta)} \in K_p[[G^{(\Delta)}]] \quad .$$

PROPOSITION 5.4.1: $\mathrm{Sm}_{r_1}^{r_2}(\mu_E^{(\Delta)})$ is a measure.

Proof: Let $M \subseteq K_p$ be the $\mathfrak{O}$-module generated by $\kappa\,\xi_E(SL_2(\mathbb{Z}))$. Then M is finitely generated. Let $g \in \mathfrak{F}((\mathbb{Q}/\mathbb{Z})_{p,\Delta}; K_p/M)$ be the reduction of $g_E^{(\Delta)}$ modulo M. We will show $(1 - r_1\,\sigma(r_1))\,(1 - r_2\,\sigma(r_2)^{-1}) \cdot g = 0$.

For $x \in (\mathbb{Q}/\mathbb{Z})_{p,\Delta}$ represented by $\frac{a}{c}$ with $(a, c) = 1$ let $x' \in (\mathbb{Q}/\mathbb{Z})_{p,\Delta}$ be the element represented by $\frac{d}{c}$ with $d \in \mathbb{Z}$ chosen so that $ad \equiv 1 \pmod{c}$. Then 2.5.3 shows that $g = g_1 + g_2$ with g_1, g_2 given by

$$g_1(x) \equiv (N_2 x \cdot a_0(E))^{\kappa} \pmod{M}$$

$$g_2(x) \equiv ((N_2 x)' \cdot a_0(E|\alpha))^{\kappa} \pmod{M}$$

where $\alpha \in SL_2(\mathbb{Z})$ satisfies $\alpha \cdot i\infty \equiv N_2 x \pmod{N_2\mathbb{Z}}$.

To complete the proof we calculate.

$$[(1 - r_1\,\sigma(r_1)) \cdot g_1](x) \equiv \{[N_2 x - r_1(\sigma(r_1)^{-1} \cdot (N_2 x))]\,a_0(E)\}^{\kappa}$$

$$\equiv 0 \pmod{M} \quad ,$$

and

$$[(1 - r_2\sigma(r_2)^{-1} \cdot g_2](x) \equiv \{[(N_2x)' - r_2 \cdot (N_2r_2x)']a_0(E|\alpha)\}^{\kappa}$$

$$\equiv 0 \pmod{M} \quad . \qquad \square$$

Let $K_p[\chi] \subseteq \mathbb{C}_p$ be the field generated over K_p by the values of χ and let

$$\mu_E(\chi) = \phi_\chi(\mu_E^{(\Delta)}) \in K_p[\chi][[U]] \quad .$$

COROLLARY 5.4.2: If $\chi, \bar{\chi}$ are nonexceptional then $\mu_E(\chi)$ is a measure on U.

Proof: Let $\mathfrak{P} \subseteq \mathfrak{O}[\chi]$ be the maximal ideal in the ring of integers of $K_p[\chi]$. Choose $r_1, r_2 \in \mathbb{Z}$ such that $(r_1r_2, p\Delta) = 1$ and $r_1\chi(r_1) \not\equiv 1$, $r_2\bar{\chi}(r_2) \not\equiv 1 \pmod{\mathfrak{P}}$. Then $(1 - r_1\chi(r_1)\langle r_1\rangle)$ and $(1 - r_2\bar{\chi}(r_2)\langle r_2\rangle)$ are invertible in $\Lambda_U(\mathfrak{O}[\chi])$. Hence

$$(1 - r_1\chi(r_1)\langle r_1\rangle)^{-1} \cdot (1 - r_2\bar{\chi}(r_2)\langle r_2\rangle)^{-1} \cdot \phi_\chi(Sm_{r_1}^{r_2}(\mu_E^{(\Delta)}))$$

$$= \mu_E(\chi) \in \Lambda_U(K_p[\chi]) \quad . \qquad \square$$

Define the p-adic L-function attached to the pair E, χ to be the function of $s \in \mathbb{Z}_p$ given by

$$L_p(E,\chi,s) = \frac{\langle\langle\ \rangle^{s-1}, \mathrm{Sm}_{r_1}^{r_2}(\mu_E(\chi))\rangle}{(1 - r_1\chi(r_1)\langle r_1\rangle^{s-1})(1 - r_2\bar{\chi}(r_2)\langle r_2\rangle^{1-s})}$$

$$\in K_p[\chi] \quad .$$

The special value at $s = 1$ is given by

$$\chi(N_2)\, L_p(E,\chi,1) =$$

$$\bar{\epsilon}_1(p^n)(1 - \bar{\epsilon}_1\chi(p))(1 - \epsilon_2\bar{\chi}(p)) \cdot \left(\frac{-\tau(\chi_\infty)\, L(E,\bar{\chi}_\infty,1)}{2\pi i}\right)^{\kappa}$$

if $\operatorname{sgn}(\chi) = \operatorname{sgn}(E)$. Otherwise it is 0.

<u>Example 5.4.3</u>: Let $\epsilon_{i,\infty} = \kappa^{-1} \circ \epsilon_i$ $(i = 1, 2)$ be the archimedean versions of ϵ_1, ϵ_2. Let $E = E(\epsilon_{1,\infty}, \epsilon_{2,\infty})$ be the Eisenstein series defined in 3.4.1 In 3.4.2 (c) we have computed $L(E,\chi_\infty, 1)$. We conclude then

$$L_p(E,\chi,1) =$$

$$\epsilon_1(\Delta)\bar{\chi}(N_1)\,[(1 - \bar{\epsilon}_1\chi(p))\, \mathbf{B}_1(\bar{\epsilon}_1\chi)] \cdot [(1 - \epsilon_2\bar{\chi}(p))\, \mathbf{B}_1(\epsilon_2\bar{\chi})] \quad .$$

Using Proposition 5.1.1 and the formulae at the end of §5.2 we find that if $\chi, \bar{\chi}$ are nonexceptional $L_p(E,\chi,s)$ can be expressed in terms of Kubota-Leopoldt p-adic L-functions:

$$L_p(E,\chi,s) =$$

$$\epsilon_1(\Delta) \cdot \bar{\chi}(N_1)\langle N_1\rangle^{1-s} \cdot L_p(\bar{\epsilon}_1\chi\omega, 1-s) \cdot L_p(\epsilon_2\bar{\chi}\omega, s-1) \quad .$$

□

§5.5. The Modular Symbol associated to E:

In this section and the next one we will assume $\epsilon_{1,\infty}, \epsilon_{2,\infty}$ are primitive and $E = E(\epsilon_{1,\infty}, \epsilon_{2,\infty})$. We also assume $p \nmid 2N$.

Let $R(E)_p$ be the completion of $R(E)^{\mathcal{K}}$ in K_p. Define

$$f_E : \mathbf{P}^1(\mathbb{Q}) \longrightarrow K_p/R(E)_p$$

$$x \longmapsto \mathcal{K}\xi_E(\gamma_x) \pmod{R(E)_p}$$

where $\gamma_x \in SL_2(\mathbf{Z})$ satisfies $\gamma_x \cdot i\infty = x$. This map is well defined: if $\alpha = \begin{pmatrix} 1 & k \\ 0 & 1 \end{pmatrix}$, $k \in \mathbf{Z}$ then $\xi_E(\gamma_x \cdot \alpha) = \xi_E(\gamma_x) + \xi_{E|\gamma_x}(\alpha)$; but $N \cdot \xi_{E|\gamma_x}(\alpha) = Nk \cdot a_0(E|\gamma_x) \in R(E)$, so

$$\mathcal{K}\xi_{E|\gamma_x}(\alpha) \equiv 0 \pmod{R(E)_p} .$$

In the next lemma we will show that f_E is a modular symbol for Γ and is an eigenfunction for the "Hecke operators".

Let $\mathrm{Symb}(\Gamma; M)$ be the group of modular symbols for Γ with values in an abelian group M. Define operators $\langle d \rangle$, $d \in (\mathbf{Z}/N\mathbf{Z})^*$, and T_ℓ, ℓ prime, on $\mathrm{Symb}(\Gamma; M)$ by:

$$(g|\langle d\rangle)(x) = g(\sigma_d \cdot x) - g(\sigma_d \cdot i\infty) ;$$

$$(g|T_\ell)(x) = \begin{cases} (g|U_\ell)(x) \\ (g|U_\ell)(x) + (g|\langle \ell\rangle)(\ell x) & \text{if } \ell \nmid N ; \end{cases}$$

where $x \in \mathbb{P}^1(\mathbb{Q})$, $\sigma_d \in SL_2(\mathbb{Z})$ satisfies

$$\sigma_d \equiv \begin{pmatrix} * & 0 \\ 0 & d \end{pmatrix} \pmod{N} ,$$

and
$$(g|U_\ell)(x) = \sum_{k=0}^{\ell-1} g\left(\frac{x + N_2 k}{\ell}\right) .$$

For the particular example $g = \underline{\text{Univ}}$ we have

$$(g|T_\ell)(x) = T_\ell \cdot g(x) ,$$

$$(g|\langle d\rangle)(x) = \langle d\rangle \cdot g(x) ,$$

where the symbols T_ℓ, $\langle d\rangle$ on the right refer to the usual action of T_ℓ, $\langle d\rangle$ on $H_1(X, \underline{\text{cusps}}; \mathbb{Z})$.

LEMMA 5.5.1: The function f_E is a modular symbol for Γ. For $x \in \mathbb{P}^1(\mathbb{Q})$ the following equalities hold:

(a) For $\gamma \in \Gamma$,

$$f_E(\gamma x) - f_E(x) \equiv \varkappa \xi_E(\gamma) \pmod{R(E)_p} ,$$

(b) $f_E(-x) = \operatorname{sgn}(E) \cdot f_E(x)$,

(c) $f_E | \langle d \rangle = \epsilon(d) \cdot f_E \quad , \quad d \in (\mathbb{Z}/N\mathbb{Z})^*$,

(d) $f_E | T_\ell = (\epsilon_1(\ell) + \ell\epsilon_2(\ell)) \cdot f_E \quad , \quad \ell$ prime .

__Proof:__ To show that f_E is a modular symbol it suffices to check (a):

$$f_E(\gamma x) - f_E(x) \equiv \kappa(\xi_E(\gamma\gamma_x) - \xi_E(\gamma_x))$$
$$= \kappa\xi_E(\gamma) \quad .$$

To verify (b) let $\iota = \begin{pmatrix} 1 & 0 \\ 0 & -1 \end{pmatrix}$. Then

$$f_E(-x) \equiv \kappa\xi_E(\iota^{-1}\gamma_x\iota) \equiv \mathrm{sgn}(E)\kappa\xi_E(\gamma_x)$$
$$\equiv \mathrm{sgn}(E) \cdot f_E(x) \quad .$$

We can also check (c) easily:

$$(f_E | \langle d \rangle)(x) = f_E(\sigma_d \cdot x) - f_E(\sigma_d \cdot i\infty)$$
$$\equiv \kappa(\xi_E(\sigma_d\gamma_x) - \xi_E(\sigma_d))$$
$$= \kappa(\xi_E |_{\sigma_d}(\gamma_x))$$
$$\equiv \epsilon(d) \cdot f_E(x) \quad .$$

The verification of (d) is more difficult. Let $W \subseteq \mathbb{P}^1(\mathbb{Q})$ be the set of rational numbers whose denominators are prime to p. We will use the following lemma.

LEMMA 5.5.2: Let $s \in \mathrm{Symb}(\Gamma; M)$. Suppose s vanishes on W and

$s(\gamma x) = s(x)$ for $x \in W$ and $\gamma \in \Gamma$. Then $s = 0$.

Proof: For each $x \in \mathbf{P}^1(\mathbb{Q})$ either $x \in W$ or $\begin{pmatrix} 1 & 0 \\ N_1 & 1 \end{pmatrix} \cdot x \in W$. In either case $s(x) = 0$. $\square$

Let $g = f_E | T_\ell$ for a prime ℓ. By the lemma, it suffices to show

$$\text{(i)}\ \ g(\gamma x) - g(x) = (\epsilon_1(\ell) + \ell\,\epsilon_2(\ell))\,\varkappa\,\xi_E(\gamma)\ ,$$

(5.5.3) $$\text{for } \gamma \in \Gamma,\ x \in \mathbf{P}^1(\mathbb{Q})\ ;$$

and $$\text{(ii)}\ \ g(x) = (\epsilon_1(\ell) + \ell\,\epsilon_2(\ell)) \cdot f_E(x) \quad \text{for} \quad x \in W\ .$$

The identity (i) follows immediately since

$$g(\gamma x) - g(x) = \varkappa\,\xi_{E|T_\ell}(\gamma)\ .$$

To prove (ii) we will use the explicit formulas of §3.4. For example, for $x \in \mathbf{P}^1(\mathbb{Q})$

(5.5.4) $$f_E(x) \equiv \begin{cases} 0 & \text{if } x = i\infty\ , \\ f_1(x) + f_2(x) - \varkappa\, s_E(x) & \text{if } x \neq i\infty\ , \end{cases}$$

where $f_1(x) = \varkappa(x \cdot a_0(E))$ and $f_2(x) = \varkappa(x' \cdot a_0(E|\gamma_x))$.

For an arbitrary prime ℓ and $x \in \mathbb{Q}$ we have

$$[(s_E|U_\ell)(x) + \varepsilon_\infty(\ell)\, s_E(\ell x)]^\kappa$$

(5.5.5)

$$= \begin{cases} \displaystyle\sum_{k=0}^{\ell-1} \tilde{e}_E\left(\begin{pmatrix}1 & N_2k\\ 0 & \ell\end{pmatrix}\begin{pmatrix}1 & x\\ 0 & 1\end{pmatrix}\right) + \tilde{e}_E|\sigma_\ell\left(\begin{pmatrix}\ell & 0\\ 0 & 1\end{pmatrix}\begin{pmatrix}1 & x\\ 0 & 1\end{pmatrix}\right) & \text{if } \ell \nmid N \\ \displaystyle\sum_{k=0}^{\ell-1} \tilde{e}_E\left(\begin{pmatrix}1 & N_2k\\ 0 & \ell\end{pmatrix}\begin{pmatrix}1 & x\\ 0 & 1\end{pmatrix}\right) & \text{if } \ell \mid N \end{cases}.$$

$$= \tilde{e}_E|T_\ell\begin{pmatrix}1 & x\\ 0 & 1\end{pmatrix} = (\varepsilon_1(\ell) + \ell\,\varepsilon_2(\ell)) \cdot \kappa s_E(x) .$$

Fix $x \in W$. Then $f_1(x) \equiv f_2(x) \equiv 0 \pmod{R(E)_p}$. Hence

$$f_E(x) = -\kappa s_E(x) .$$

Now suppose $\ell \neq p$. Then $\ell x \in W$ and $\dfrac{x + N_2 k}{\ell} \in W$ for $k \in \mathbb{Z}$. Hence by (5.5.5)

$$g(x) = (f_E|U_\ell)(x) + \varepsilon(\ell) \cdot f_E(\ell x)$$

$$= (\varepsilon_1(\ell) + \ell\,\varepsilon_2(\ell)) \cdot f_E(x) ,$$

as desired.

Finally, consider the case $\ell = p$. We will verify the formula

(5.5.6) $$(s|U_p)(x) + \varepsilon(p)\, s(px) = (\varepsilon_1(p) + p\,\varepsilon_2(p))\, s(x)$$

for $x \in \mathbb{Q}$ and $s = f_1$ or $s = f_2$. Since we have already checked this for $s = \kappa s_E$, (5.5.6) will hold for $s = f_E$ as well which is the desired result.

If $s = f_1$ the left-hand side of (5.5.6) is easily seen to be $(1 + p\epsilon(p)) \cdot f_1(x)$. But $a_0(E) = 0$ unless $N_1 = 1$ (Theorem 3.4.4 c) so this is equal to $(\epsilon_1(p) + p\epsilon_2(p))\, f_1(x)$.

Suppose $x = \frac{a}{c}$ with $c > 0$. Since $x \in W$, we know $p \nmid c$. Let $a', p' \in \mathbb{Z}$ be such that $a a' \equiv p p' \equiv 1 \pmod{c}$. For each $k \in \mathbb{Z}/p\mathbb{Z}$ there is a unique $r(k) \in \mathbb{Z}/p\mathbb{Z}$ such that $\left(\frac{a + N_2 c k}{pc}\right)' = \frac{a' + N_2 c \cdot r(k)}{pc}$.

The map $r : \mathbb{Z}/p\mathbb{Z} \to \mathbb{Z}/p\mathbb{Z}$ is a bijection. Since $p \nmid N_2 c$ there is a unique $k_0 \in \mathbb{Z}/p\mathbb{Z}$ for which $a + N_2 c k_0 \equiv 0 \pmod{p}$.

Using Theorem 3.4.4 c we find

$$f_2\left(\frac{x + N_2 k_0}{p}\right) = f_2(p'x) = p\epsilon_2(p) \cdot f_2(x) \quad ;$$

$$f_2\left(\frac{x + N_2 k}{p}\right) \equiv \frac{a' + N_2 c \cdot r(k)}{pc} \cdot \epsilon_1(p) \cdot \mathcal{K}(a_0(E|\gamma_x)) \quad \text{if } k \neq k_0 \; ;$$

and

$$\epsilon(p) \cdot f_2(px) \equiv \epsilon(p) \cdot \frac{p' a'}{c} \cdot \bar{\epsilon}_2(p) \cdot \mathcal{K}(a_0(E|\gamma_x))$$

$$\equiv \frac{a' + N_2 c \cdot r(k_0)}{pc} \cdot \epsilon_1(p) \cdot \mathcal{K}(a_0(E|\gamma_x)) \quad .$$

Summing these yields (5.5.6) with $s = f_2$. □

§5.6. Congruences Between p-adic L-functions:

In this section we will use f_E to construct intermediate measures $\lambda_E(\chi)$ for the purposes of proving "congruences" between the universal measures $\mu_P(\chi)$ and the measures $\mu_E(\chi)$ where χ ranges through a set of nonexceptional $\mathbb{C}_p$-valued Dirichlet characters. If f is a weight two cusp form satisfying conditions like those of (4.2.2) we prove a congruence between the measures $\mu_f(\chi)$ and $\mu_E(\chi)$.

Let $\mathfrak{O} = \mathbb{Z}_p[\epsilon_1, \epsilon_2] \subseteq \mathbf{D}_p$ be the discrete valuation ring generated over $\mathbb{Z}_p$ by the values of ϵ_1, ϵ_2. Let Periods $(E)_p$ be the completion of Periods $(E)^{\kappa}$ in K_p. Since $p \nmid 2N$, Theorem 3.6.1 tells us

$$\text{Periods}\,(E)_p = \mathfrak{O} \quad .$$

Then $R(E)_p$ is an integral ideal of $\mathfrak{O}$. Let

$$A(E)_p = \mathfrak{O}/R(E)_p \quad .$$

Suppose $A(E)_p \neq 0$.

Let $\mathbb{M}_p$ be the maximal ideal of $\mathbf{D}_p$ and $\mathfrak{m} \subseteq \mathfrak{J}$ be the pullback of $\mathbb{M}_p$ to $\mathfrak{J}$ under $\kappa \circ h_E : \mathfrak{J} \to \mathbf{D}_p$. Let $P \subseteq \mathbb{T}$ be the image of $\mathfrak{m}$. Then P is an Eisenstein prime associated to E. There is a natural $\mathbb{T}$-isomorphism

$$A(E)_p \cong A(E) \otimes_{\mathbb{T}} \mathbb{T}_P \quad .$$

We will need a few lemmas.

LEMMA 5.6.1: The inclusion

$$H_P^{\mathrm{sgn}(E)} \hookrightarrow H_P^{\partial,\,\mathrm{sgn}(E)}$$

is an isomorphism.

Proof: Let $B = H_0(\underline{\text{cusps}};\mathbb{Z})$. Consider the exact sequence

$$0 \longrightarrow H_P^{\mathrm{sgn}(E)} \xrightarrow[j_P]{} H_1(X,\underline{\text{cusps}};\mathbb{Z})_{\mathfrak{m}}^{\mathrm{sgn}(E)} \longrightarrow B_{\mathfrak{m}}^{\mathrm{sgn}(E)} \quad .$$

By (3.2.3), if $\ell \equiv -1 \pmod N$ is a prime, then

$$T_\ell - \epsilon_1(-1)(\ell+1) \in \mathfrak{I}(E) \subseteq \mathfrak{m} \quad .$$

Hence T_ℓ acts by multiplication by $\epsilon_1(-1)(\ell+1)$ on $B/\mathfrak{m}\cdot B$. On the other hand, the action of T_ℓ on a cusp $[{x \atop y}]$ is given by

$$\begin{aligned} T_\ell \cdot [{x \atop y}] &= \ell \cdot [{x \atop \ell y}] + [{\ell' x \atop y}] \\ &= (\ell+1)\cdot \iota([{x \atop y}]) \quad . \end{aligned}$$

Hence

$$(\ell+1)(\iota - \epsilon_1(-1)) \cdot B \subseteq \mathfrak{m}\cdot B \quad .$$

Since this is true for every prime $\ell \equiv -1 \pmod N$ we must have $2N \cdot (\iota - \epsilon_1(-1)) \cdot B \subseteq \mathfrak{m}\cdot B$. But $2N \notin \mathfrak{m}$ and $\mathrm{sgn}(E) = -\epsilon_1(-1)$, so

$$(\iota + \mathrm{sgn}(E)) \cdot B \subseteq \mathfrak{m}\cdot B \quad .$$

Therefore $(\iota - \mathrm{sgn}(E)) \cdot B + \mathfrak{m} \cdot B = B$. By Nakayama's lemma $B_{\mathfrak{m}}^{\mathrm{sgn}(E)} = 0$. So j_P is an isomorphsim.

The homomorphism $\eta_P^{\mathrm{sgn}(E)} : H_1(X, \text{cusps}; \mathbb{Q})_{\mathfrak{m}}^{\mathrm{sgn}(E)} \longrightarrow H_1(X; \mathbb{Q})_P^{\mathrm{sgn}(E)}$ is the unique $\mathfrak{T}_{\mathfrak{m}}$-homomorphism for which $\eta_P^{\mathrm{sgn}(E)} \circ j_P$ is the identity. Hence $\eta_P^{\mathrm{sgn}(E)} = j_P^{-1}$, and $H_P^{\mathrm{sgn}(E)} = H_P^{\partial, \mathrm{sgn}(E)}$. □

By Theorem 4.1.6 P is an ordinary prime. We may therefore proceed as in §5.3 and construct a universal measure $\mu_P^{(\Delta), \mathrm{sgn}(E)} \in H_P^{\mathrm{sgn}(E)}[[G^{(\Delta)}]]$ for each $\Delta > 0$ satisfying $(\Delta, pN) = 1$. Let $g_P : \mathbf{P}^1(\mathbb{Q}) \to H_P^{\partial}$ be as in §5.3 and $g_P^{\mathrm{sgn}(E)}$ be the composition

$$g_P^{\mathrm{sgn}(E)} : \mathbf{P}^1(\mathbb{Q}) \xrightarrow{g_P} H_P^{\partial} \longrightarrow H_P^{\partial, \mathrm{sgn}(E)} \xrightarrow{\sim} H_P^{\mathrm{sgn}(E)} .$$

Let $\varphi_{E,p} : H_P^{\mathrm{sgn}(E)} \to A(E)_p$ be the completion at P of the homomorphism $\varphi_E^{\mathrm{sgn}(E)}$.

LEMMA 5.6.2: The modular symbol f_E has values in $A(E)_p$, and the following diagram is commutative:

$$\begin{array}{ccc} \mathbb{P}^1(\mathbb{Q}) & \xrightarrow{f_E} & A(E)_p \\ {\scriptstyle g_p^{\mathrm{sgn}(E)}} \big\downarrow & \nearrow {\scriptstyle \varphi_{E,p}} & \\ H_p^{\mathrm{sgn}(E)} & & \end{array}$$

<u>Proof</u>: By Lemma 5.5.1 there is a $\mathfrak{J}$-homomorphism $\phi : H_1(X, \underline{\text{cusps}}; \mathbb{Z})^{\mathrm{sgn}(E)} \to K_p/R(E)_p$ satisfying $\phi \circ \mathrm{Univ}^{\mathrm{sgn}(E)} = f_E$. Let $\phi_{\mathfrak{m}}$ be the completion of ϕ at $\mathfrak{m}$. If $j_p : H_p^{\mathrm{sgn}(E)} \xrightarrow{\sim} H_1(X, \underline{\text{cusps}}; \mathbb{Z})_{\mathfrak{m}}^{\mathrm{sgn}(E)}$ as in the proof of Lemma 5.6.1, then

$$j_p^{-1} \circ \underline{\mathrm{Univ}}_{\mathfrak{m}}^{\mathrm{sgn}(E)} = g_p^{\mathrm{sgn}(E)} ,$$

and

$$\phi_{\mathfrak{m}} \circ j_p = \varphi_{E,p} .$$

So the image of f_E is contained in the image of $\varphi_{E,p}$ and $f_E = \varphi_{E,p} \circ g_p^{\mathrm{sgn}(E)}$. □

For each $\Delta > 0$ satisfying $(\Delta, pN) = 1$, we can apply (II) of §5.2 to f_E and construct a measure $\lambda_E^{(\Delta)} \in \Lambda_{G^{(\Delta)}}(A(E)_p)$. Then by Lemma 5.6.2

$$\lambda_E^{(\Delta)} = \varphi_{E,p}^* \cdot \mu_p^{(\Delta),\mathrm{sgn}(E)} . \tag{5.6.3}$$

Next we compare $\lambda_E^{(\Delta)}$ to the distribution $\mu_E^{(\Delta)}$.

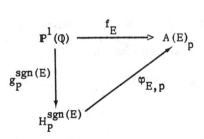

LEMMA 5.6.4: Let $r_1, r_2 \in \mathbb{Z}$ with $(r_1 r_2, p\Delta) = 1$. Then

$$\mathrm{Sm}_{r_1}^{r_2}(\mu_E^{(\Delta)}) \in \Lambda_{G^{(\Delta)}}(\mathfrak{G}) \quad .$$

Let $\pi : \mathfrak{G} \to A(E)_p$ be the natural projection. Then

$$\pi^* \mathrm{Sm}_{r_1}^{r_2}(\mu_E^{(\Delta)}) = \mathrm{Sm}_{r_1}^{r_2}(\lambda_E^{(\Delta)}) \quad .$$

<u>Proof</u>: Define $\bar{g}_E^{(\Delta)} \in \mathfrak{F}((\mathbb{Q}/\mathbb{Z})_{p,\Delta}; A(E)_p)$ by $\bar{g}_E^{(\Delta)}(x) = f_E(N_2 x)$. By (5.5.4) we have $\bar{g}_E^{(\Delta)} = (f_1 \circ N_2) + (f_2 \circ N_2) - \pi \circ g_E^{(\Delta)}$. It suffices to check the following identities

$$0 = (1 - r_1 \sigma(r_1)) \cdot (f_1 \circ N_2) = (1 - r_2 \sigma(r_2)^{-1}) \cdot (f_2 \circ N_2) \quad .$$

This is a simple calculation. □

For a Dirichlet character $\chi : \mathbb{Z} \to \mathbf{D}_p$ let $A(E)_p[\chi] =$
$= \mathfrak{G}[\chi]/(R(E)_p \cdot \mathfrak{G}[\chi])$ and let

$$\pi \otimes 1 : \mathfrak{G}[\chi] \longrightarrow A(E)_p[\chi]$$

and

$$(\varphi_{E,p} \otimes 1) : H_p^{\mathrm{sgn}(E)}[\chi] \longrightarrow A(E)_p[\chi]$$

be the natural homomorphisms.

THEOREM 5.6.5: Let $\chi : \mathbb{Z} \to \mathbf{D}_p$ be a nontrivial Dirichlet character of conductor prime to N. Suppose $\mathrm{sgn}(\chi) = \mathrm{sgn}(E)$ and $\chi, \bar{\chi}$ are nonexceptional. Then

i) $\mu_E(\chi) \in \Lambda_U(\mathcal{O}[\chi])$;

and ii) $(\pi \otimes 1)^* \cdot \mu_E(\chi) = (\varphi_{E,p} \otimes 1)^* \cdot \mu_p(\chi) \in \Lambda_U(A(E)_p[\chi])$.

Proof: Let $r_1, r_2 \in \mathbb{Z}$ such that $(1 - r_1\chi(r_1))$ and $(1 - r_2\bar{\chi}(r_2))$ are units in D_p. Then by Lemma 5.6.4

$$\mu_E(\chi) = (1 - r_1\chi(r_1)\langle r_1\rangle)^{-1} \cdot (1 - r_2\bar{\chi}(r_2)\langle r_2\rangle)^{-1} \cdot \phi_\chi\left(\mathrm{Sm}_{r_1}^{r_2}(\mu_E^{(\Delta)})\right)$$

$$\in \Lambda_U(\mathcal{O}[\chi]) \quad .$$

Lemma 5.6.4 together with (5.6.3) show:

$$(\pi \otimes 1)^* \cdot \phi_\chi\left(\mathrm{Sm}_{r_1}^{r_2}(\mu_E^{(\Delta)})\right) = \phi_\chi \cdot \pi^* \cdot \mathrm{Sm}_{r_1}^{r_2}(\mu_E^{(\Delta)})$$

$$= \phi_\chi \cdot \mathrm{Sm}_{r_1}^{r_2}(\lambda_E^{(\Delta)})$$

$$= \phi_\chi \cdot \varphi_{E,p}^* \cdot \mathrm{Sm}_{r_1}^{r_2}(\mu_p^{(\Delta),\,\mathrm{sgn}(E)})$$

$$= (\varphi_{E,p} \otimes 1)^* \cdot \phi_\chi\left(\mathrm{Sm}_{r_1}^{r_2}(\mu_p^{(\Delta),\,\mathrm{sgn}(E)})\right) \quad .$$

Applying $(1 - r_1\chi(r_1)\langle r_1\rangle)^{-1} \cdot (1 - r_2\bar{\chi}(r_2)\langle r_2\rangle)^{-1}$ to both sides of this equality yields the desired results. □

COROLLARY 5.6.6: For $s \in \mathbb{Z}_p$, and χ as above,

$$(\varphi_{E,p} \otimes 1) \cdot L_p(X, \chi, s) \equiv$$

$$\epsilon_1(\Delta) \cdot \overline{\chi}(N_1) \cdot \langle N_1 \rangle^{1-s} \cdot L_p(\overline{\epsilon}_1 \chi\omega, 1-s) \cdot L_p(\epsilon_2 \overline{\chi}\omega, s-1)$$

$$(\text{mod } R(E)_p \cdot \mathfrak{O}[\chi]) \quad .$$

Proof: This is immediate from the theorem and the calculation of Example 5.4.3. □

Now let $f \in \mathcal{S}_2(\Gamma)$ be a cusp form for Γ which is an eigenfunction for $\mathbb{T}$. Let $\mathfrak{B} \subseteq \overline{\mathbb{Z}}$ be the pullback of $\mathbb{M}_p$ under the homomorphism $\kappa : \overline{\mathbb{Z}} \to \mathbb{D}_p$, and let $\mathfrak{P} = \mathfrak{B} \cap \mathfrak{O}(f)$.

Suppose $\mathfrak{P}$ satisfies the conditions (4.2.2). Without loss of generality we may assume $E = E_\sigma$ where $\sigma \in \mathrm{Gal}(\overline{\mathbb{Q}}/\mathbb{Q})$ is as in Theorem 4.2.3. Let $\theta_p : \mathbb{D}_p \to \overline{\mathbb{F}}_p \cong \mathbb{D}_p / \mathbb{M}_p$ be the natural projection.

THEOREM 5.6.7: With the assumptions and notations of the last two paragraphs, there is a period $\Omega_p \in \mathbb{C}_p^*$ such that

(i) $\frac{1}{\Omega_p} \cdot \mu_f(\chi) \in \Lambda_U(\mathbb{D}_p)$;

and (ii) $\theta_p^* \cdot \left(\frac{1}{\Omega_p} \cdot \mu_f(\chi)\right) = \theta_p^* \cdot \mu_E(\chi) \in \Lambda_U(\overline{\mathbb{F}}_p)$;

for every nontrivial primitive Dirichlet character χ of conductor prime to N, for which $\mathrm{sgn}(\chi) = \mathrm{sgn}(E)$ and $\chi, \overline{\chi}$ are nonexceptional.

Proof: We will use the notation of the proof of Theorem 4.2.3.

We may choose $\Omega_E \in \underline{\text{Periods}}(E)$ such that $\varkappa(\Omega_E) \equiv 1 \pmod{\mathbb{M}_p}$. Let $\Omega_f \in \underline{\text{Periods}}(f)^{\mathrm{sgn}(E)}$ be such that the image of Ω_f in $A(E)_p$ is $1 \pmod{R(E)_p}$, and set $\Omega_p = \varkappa(\Omega_f) \in \mathbb{C}_p^*$. Let $\varphi_{f,p}$ be the P-adic completion of $\kappa \circ \varphi_f^{\mathrm{sgn}(E)} : H^{\mathrm{sgn}(E)} \to A(E)_p$. The following diagram is commutative:

$$\begin{array}{ccccc}
H_p^{\mathrm{sgn}(E)} & \xrightarrow{\ \varphi_{E,p}\ } & A(E)_p & \xleftarrow{\ \pi\ } & \mathcal{O} \\
\Big\downarrow{\scriptstyle \frac{1}{\Omega_p}\cdot\varphi_{f,p}} & & \Big\downarrow{\scriptstyle \bar{\theta}_p} & \swarrow{\scriptstyle \theta_p} & \\
\mathbb{D}_p & \xrightarrow[\ \theta_p\ ]{} & \bar{\mathbb{F}}_p & &
\end{array}$$

For $\Delta > 0$ with $(\Delta, pN) = 1$, choose $r_1, r_2 \in \mathbb{Z}$ such that $(r_1 r_2, p\Delta) = 1$, and $(1 - r_1\chi(r_1))$, $(1 - r_2\bar{\chi}(r_2))$ are units in $\mathbb{D}_p$. Then by Lemma 5.6.4, (5.6.3)

$$\begin{aligned}
\theta_p^* \cdot \mathrm{Sm}_{r_1}^{r_2}(\mu_E^{(\Delta)}) &= \bar{\theta}_p^* \cdot \pi^* \cdot \mathrm{Sm}_{r_1}^{r_2}(\mu_E^{(\Delta)}) \\
&= \bar{\theta}_p^* \cdot \varphi_{E,p}^* \cdot \mathrm{Sm}_{r_1}^{r_2}(\mu_p^{(\Delta),\mathrm{sgn}(E)}) \\
&= \theta_p^*(\Omega_p^{-1}\, \varphi_{f,p}^*\, \mathrm{Sm}_{r_1}^{r_2}(\mu_p^{(\Delta),\mathrm{sgn}(E)})) \\
&= \theta_p^*(\Omega_p^{-1} \cdot \mathrm{Sm}_{r_1}^{r_2}(\mu_f^{(\Delta),\mathrm{sgn}(E)})) .
\end{aligned}$$

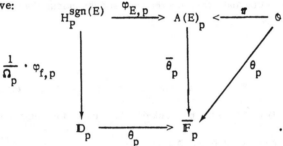

Now just apply $(1 - r_1 \chi(r_1) \langle r_1 \rangle)^{-1} \cdot (1 - r_2 \bar{\chi}(r_2) \langle r_2 \rangle)^{-1} \cdot \phi_\chi$ to both sides of this equation. □

In terms of p-adic L-function this theorem can be expressed as follows.

COROLLARY 5.6.8: With the assumptions of the theorem, we have for $s \in \mathbb{Z}_p$

$$\Omega_p^{-1} \cdot L_p(f, \chi, s) \equiv$$

$$\epsilon_1(\Delta) \cdot \bar{\chi}(N_1) \cdot \langle N_1 \rangle^{1-s} \cdot L_p(\bar{\epsilon}_1 \chi\omega, 1-s) \cdot L_p(\epsilon_2 \bar{\chi}\omega, s-1)$$

$$(\text{mod } \mathbb{M}_p) \quad . \qquad \square$$

Chapter 6. Tables of Special Values

In the remaining pages we display three sets of tables of algebraic parts of special values of L-functions,

$$\Lambda(\chi) = \frac{\tau(\bar{\chi}) \cdot L(f, \chi, 1)}{\Omega_f^{\mathrm{sgn}(\chi)}} .$$

Here χ denotes a primitive quadratic character of conductor m_χ. In the first two sets of tables m_χ is taken to be positive or negative depending on whether $\chi(-1) = \mathrm{sgn}\,\chi$ is plus or minus one. The modular form f ranges through the weight two parabolic eigenforms for the following modular curves:

1. $X_0(N)$, N prime ≤ 43 ;
2. Genus one curves $X_0(N)$,

   $N = 14, 15, 20, 21, 24, 27, 32, 36, 49$;
3. $X_1(13)$.

The complex number $\Omega_f^{\mathrm{sgn}\,\chi}$ is an appropriate period of $f(z)\,dz$ on the corresponding modular curves.

These tables have been obtained using an algorithm due to Birch [2], and Manin [25].

§6.1 $X_0(N)$, N prime ≤ 43

Let f be a weight two parabolic eigenform for $\Gamma_0(N)$, N prime. By a recent result of Waldspurger [44] there are constants $\alpha^{\pm} \in K(f)$ such that for all primitive quadratic characters χ with conductor m_χ prime to N the special value

$$\Lambda(\chi) = \frac{\tau(\bar{\chi})\, L(f, \chi, 1)}{\Omega_f^{\operatorname{sgn}\chi}}$$

and be expressed as $\Lambda(\chi) = \alpha^{\operatorname{sgn}\chi} \cdot \psi(\chi)^2$ with $\psi(\chi) \in K(f)$.

In the following tables we give not $\Lambda(\chi)$ but $\psi(\chi)$ for such characters with $|m_\chi| < 500$. The relationship between $\Lambda(\chi)$ and $\psi(\chi)$ is printed at the top of each page.

Of course the sign of $\psi(\chi)$ is not uniquely determined. In case $\chi(-1) = -1$ we have chosen $\psi(\chi)$ to be congruent to a fixed multiple of $B_1(\chi)$ modulo the Eisenstein ideal $I \subseteq \mathcal{O}(f)$. We know there is such a choice by a theorem of B. Mazur [22]. The Hecke algebras and Eisenstein ideals corresponding to our choice of f in the following tables are:

$X_0(11)$, $\mathbb{T}_{\mathbb{Q}} = \mathbb{Q}$

$I = 5\mathbb{Z}$,

$X_0(17)$, $\mathbb{T}_{\mathbb{Q}} = \mathbb{Q}$

$I = 4\mathbb{Z}$,

$X_0(19)$, $\mathbb{T}_{\mathbb{Q}} = \mathbb{Q}$

$I = 3\mathbb{Z}$,

$X_0(23)$, $\mathbb{T}_{\mathbb{Q}} = \mathbb{Q}(\omega)$, $\omega = \frac{1+\sqrt{5}}{2}$,

$I = (3+\omega)\cdot\mathbb{Z}[\omega]$, $N(I) = 11$;

$X_0(29)$, $\mathbb{T}_{\mathbb{Q}} = \mathbb{Q}(\sqrt{2})$,

$I = (3-\sqrt{2})\cdot\mathbb{Z}[\sqrt{2}]$, $N(I) = 7$;

$X_0(31)$, $\mathbb{T}_{\mathbb{Q}} = \mathbb{Q}(\omega)$, $\omega = \frac{1+\sqrt{5}}{2}$,

$I = (1-2\omega)\cdot\mathbb{Z}[\omega]$, $N(I) = 5$;

$X_0(37)$, $\mathbb{T}_{\mathbb{Q}} = \mathbb{Q}\times\mathbb{Q}$,

$I = \mathbb{Z}\times 3\mathbb{Z}$;

$X_0(41)$, $\mathbb{T}_{\mathbb{Q}} = \mathbb{Q}(\omega)$, $\omega^3 - 4\omega + 2 = 0$,

$I = (4-7\omega-\omega^2)\cdot\mathbb{Z}[\omega]$, $N(I) = 10$;

$X_0(43)$, $\mathbb{T}_{\mathbb{Q}} = \mathbb{Q}\times\mathbb{Q}(\sqrt{2})$,

$I = (-1+2\sqrt{2})\cdot\mathbb{Z}[\sqrt{2}]$, $N(I) = 7$.

In the following tables, the absence of an entry next to a conductor m_χ indicates a trivial zero; that is, $\Lambda(\chi) = 0$ because of the sign of the functional equation.

If $N|m_\chi$, the symbol - appears in the $\psi(\chi)$ column.

As an example we read a few entries from the tables for $X_0(23)$.

If χ is the Legendre symbol $(\frac{-}{29})$ we read from the table on page 177. We find $\psi(\chi) = 3 - 2\omega$ where $\omega = (1+\sqrt{5})/2$. Reading from the top of the page we see

$$\Lambda(\chi) = (3+\omega)(3-2\omega)^2 = 31 - 19\omega \quad .$$

If $\chi = (\frac{-}{31})$ we read from page 176

$$\Lambda(\chi) = (3-2\omega)^2 = 13 - 8\omega \quad .$$

The square root $3 - 2\omega$ has been chosen to be congruent modulo $3+\omega$ to $3 \cdot h$ where $h = 3$ is the class number of $\mathbb{Q}(\sqrt{-31})$.

As a last example, if $\chi = (\frac{-}{11})$ then $\Lambda(\chi) = 0$ because of the sign of the functional equation.

$X_0(11)$: Odd quadratic characters

$\Lambda(\chi) = a^2$

m_χ	a	m_χ	a	m_χ	a	m_χ	a	m_χ	a
-3	1	-4	-1	-7		-8		-11	-
-15	1	-19		-20	1	-23	-1	-24	
-31	-1	-35		-39		-40		-43	
-47	0	-51		-52		-55	-	-56	2
-59	-1	-67	3	-68		-71	1	-79	
-83		-84		-87		-88	-	-91	-4
-95		-103	0	-104	-2	-107		-111	-1
-115	1	-116		-119	0	-120		-123	
-127		-131		-132	-	-136	2	-139	
-143	-	-148	1	-151		-152	-2	-155	-3
-159	0	-163	-2	-164		-167		-168	2
-179	5	-183		-184		-187	-	-191	-1
-195		-199	2	-203	2	-211		-212	-2
-215		-219		-223	1	-227		-228	
-231	-	-232	-4	-235	6	-239		-244	
-247	-2	-248		-251	1	-255		-259	
-260		-263		-264		-267	1	-271	
-276	-1	-280	2	-283		-287	2	-291	-3
-292		-295	-1	-296		-299		-303	
-307		-308	-	-311	2	-312	2	-319	-
-323	2	-327		-328	2	-331	-1	-335	-1
-339	-7	-340		-344	0	-347		-355	-3
-356	1	-359		-367	-3	-371		-372	-3
-376		-379	-1	-383	1	-388	-3	-391	
-395		-399	-2	-403		-404		-407	-
-408	2	-411	3	-415		-419	2	-420	
-424		-427	6	-431		-435		-436	
-439		-440	-	-443	5	-447		-451	-
-452	-1	-455	0	-456	-6	-463	1	-467	1
-471	3	-472		-479		-483		-487	1
-488	0	-491		-499	-6				

$X_0(11)$: Even quadratic characters

$$\Lambda(\chi) = 5 \cdot a^2$$

m_χ	a	m_χ	a	m_χ	a	m_χ	a	m_χ	a
5	1	8		12	1	13		17	
21		24		28		29		33	-
37	1	40		41		44	-	53	2
56	2	57		60	1	61		65	
69	3	73		76		77	-	85	
88	-	89	1	92	1	93	1	97	1
101		104	2	105		109		113	1
120		124	1	129		133	2	136	2
137	1	140		141	2	145		149	
152	2	156		157	3	161		165	-
168	2	172		173		177	1	181	3
184		185	1	188	2	193		197	
201	1	204		205		209	-	213	1
217		220	-	221	2	229	1	232	0
233		236	1	237		241		248	
249		253	-	257	2	264	-	265	0
268	1	269	4	273	0	277		280	2
281		284	3	285		293		296	
301	2	305		309	4	312	2	313	1
316		317	1	321		328	2	329	
332		337		341	-	344	0	345	1
348		349		353	1	357	4	364	0
365		373		376		377	2	380	
381		385	-	389	1	393		397	2
401	0	408	2	409		412	2	413	
417		421	0	424		428		429	-
433	1	437		440	-	444	3	445	1
449	1	453		456	2	457		460	1
461		465	3	469		472		473	-
476	0	481		485	5	488	0	489	2
492		493	2	497					

$X_0(17)$: Odd quadratic characters

$\Lambda(\chi) = a^2$

m_χ	a	m_χ	a	m_χ	a	m_χ	a	m_χ	a
-3	1	-4		-7	-1	-8		-11	-1
-15		-19		-20	2	-23	1	-24	2
-31	1	-35		-39	0	-40	2	-43	
-47		-51	-	-52		-55		-56	0
-59		-67		-68	-	-71	1	-79	-1
-83		-84		-87		-88	2	-91	2
-95	0	-103		-104		-107	1	-111	
-115		-116	2	-119	-	-120		-123	
-127		-131	-1	-132		-136	-	-139	-3
-143	2	-148	2	-151		-152		-155	
-159	2	-163	-1	-164	0	-167	-3	-168	
-179		-183		-184	0	-187	-	-191	
-195		-199	3	-203		-211	1	-212	
-215	2	-219		-223		-227	-5	-228	4
-231	0	-232	2	-235	2	-239		-244	2
-247		-248	4	-251		-255	-	-259	
-260	0	-263		-264		-267	2	-271	
-276		-280		-283	-3	-287		-291	
-292	4	-295	0	-296	2	-299	4	-303	2
-307		-308		-311	-3	-312	4	-319	
-323	-	-327		-328	4	-331		-335	2
-339		-340	-	-344		-347	3	-355	
-356		-359		-367	-1	-371	0	-372	
-376		-379	1	-383		-388	4	-391	-
-395		-399		-403	2	-404		-407	
-408	-	-411	2	-415	2	-419	3	-420	0
-424		-427		-431	-1	-435	4	-436	2
-439	1	-440		-443		-447	2	-451	
-452	0	-455		-456	0	-463		-467	
-471	0	-472		-479	3	-483	0	-487	1
-488	6	-491		-499	1				

$X_0(17)$: Even quadratic characters

$$\Lambda(\chi) = 4 \cdot a^2$$

m_χ	a	m_χ	a	m_χ	a	m_χ	a	m_χ	a
5		8	1	12		13	1	17	-
21	1	24		28		29		33	1
37		40		41		44		53	2
56		57		60	2	61		65	
69	1	73		76	0	77	1	85	-
88		89	1	92		93	3	97	
101	1	104	0	105		109		113	
120	2	124		129		133		136	-
137	1	140	2	141		145	0	149	2
152	2	156		157	0	161	1	165	
168	2	172	2	173		177		181	
184		185	0	188	2	193		197	
201		204	-	205	2	209		213	1
217	1	220	2	221	-	229	1	232	
233		236	2	237	1	241		248	
249		253	3	257	1	264	0	265	
268	4	269		273	2	277		280	2
281	2	284		285	0	293	0	296	
301		305	2	309		312		313	
316		317		321	1	328		329	
332	2	337		341	1	344	0	345	
348	2	349	0	353	0	357	-	364	
365	2	373	3	376	2	377		380	
381		385		389	1	393	3	397	
401		408	-	409	0	412	0	413	
417	3	421	3	424	2	428		429	2
433	1	437		440	2	444	2	445	
449		453		456		457	3	460	2
461	0	465		469		472	0	473	
476	-	481		485	4	488		489	1
492	4	493	-	497	1				

$\chi_0(19)$: Odd quadratic characters

$\Lambda(\chi) = a^2$

m_χ	a	m_χ	a	m_χ	a	m_χ	a	m_χ	a
-3		-4	1	-7	-1	-8		-11	-1
-15		-19	-	-20	1	-23	0	-24	-2
-31		-35	1	-39	2	-40		-43	-1
-47	1	-51		-52		-55	-1	-56	
-59		-67		-68	-1	-71		-79	
-83	0	-84		-87	0	-88		-91	
-95	-	-103		-104	0	-107		-111	-2
-115	-2	-116		-119	-1	-120	2	-123	-2
-127		-131	1	-132		-136		-139	3
-143		-148		-151		-152	-	-155	
-159	2	-163	2	-164		-167		-168	2
-179		-183		-184		-187	1	-191	-1
-195	2	-199	3	-203		-211		-212	
-215	1	-219		-223		-227		-228	-
-231		-232	-2	-235	1	-239	3	-244	3
-247	-	-248	-2	-251	-1	-255		-259	
-260		-263	-1	-264	-2	-267	-2	-271	4
-276		-280		-283	3	-287		-291	2
-292	-1	-295		-296	2	-299		-303	
-307		-308	1	-311	-1	-312		-319	
-323	-	-327	0	-328	2	-331		-335	
-339	0	-340	5	-344		-347	1	-355	
-356		-359	-1	-367	0	-371		-372	2
-376		-379		-383		-388		-391	-2
-395		-399	-	-403	-2	-404	-2	-407	
-408	2	-411		-415	2	-419	0	-420	
-424	6	-427	1	-431		-435	-4	-436	
-439		-440		-443	1	-447		-451	
-452		-455		-456	-	-463	-1	-467	-1
-471		-472	0	-479	2	-483		-487	
-488		-491	0	-499	3				

$X_0(19)$: Even quadratic characters

$$\Lambda(X) = 3\cdot a^2$$

m_X	a	m_X	a	m_X	a	m_X	a	m_X	a
5	1	8		12		13		17	1
21		24	2	28	1	29		33	
37		40		41		44	3	53	
56		57	-	60		61	1	65	
69		73	1	76	-	77	1	85	1
88		89		92	2	93	2	97	
101	2	104	4	105		109	-	113	
120	2	124		129		133		136	
137	1	140	1	141		145		149	3
152	-	156	2	157	0	161	4	165	
168	2	172	1	173		177	2	181	
184		185		188	3	193		197	2
201	4	204		205		209	-	213	2
217		220	1	221		229	3	232	2
233	1	236		237	0	241		248	2
249		253	0	257		264	2	265	
268		269		273	2	277	1	280	
281		284		285	-	293		296	2
301	3	305	1	309	4	312		313	2
316		317		321	0	328	2	329	3
332	2	337		341		344		345	
348	0	349	1	353	2	357		364	
365	3	373		376		377	2	380	-
381	4	385	1	389	1	393		397	1
401		408	2	409		412		413	
417		421		424	2	428		429	2
433		437	-	440		444	6	445	
449		453	2	456		457	1	460	4
461	3	465	2	469		472	0	473	3
476	7	481	2	485		488		489	
492	2	493		497					

$X_0(23)$: Odd quadratic characters

$\Lambda(\chi) = (a+b\omega)^2$; $\omega = (1+\sqrt{5})/2$

m_χ	a + b·(ω)		m_χ	a + b·(ω)		m_χ	a + b·(ω)	
-3	1	0	-4	-1	1	-7		
-8	0	-1	-11			-15		
-19			-20			-23	-	-
-24	-2	1	-31	3	-2	-35	0	-2
-39	1	0	-40			-43		
-47	-1	2	-51			-52	3	-1
-55	-4	2	-56			-59	-2	0
-67			-68			-71	-1	0
-79			-83			-84		
-87	1	-2	-88			-91		
-95	2	0	-103			-104	4	-1
-107			-111			-115	-	-
-116	-1	1	-119	2	-2	-120		
-123	1	2	-127	-1	2	-131	-1	2
-132			-136			-139	3	-2
-143			-148			-151	-1	0
-152			-155			-159		
-163	3	0	-164	-1	-1	-167	0	0
-168			-179	-1	2	-183		
-184	-	-	-187	0	-2	-191		
-195			-199			-203		
-211	-2	0	-212			-215	-2	0
-219	1	0	-223	4	-2	-227		
-228			-231	-2	2	-232	-2	1
-235			-239	1	0	-244		
-247			-248	0	3	-251		
-255	-2	2	-259	-4	2	-260		
-263			-264			-267		
-271	0	0	-276	-	-	-280	6	-2
-283			-287			-291		
-292	3	-3	-295			-296		
-299	-	-	-303	2	-2	-307	-2	0
-308	-4	-2	-311	-3	2	-312	-2	-1
-319			-323	0	-4	-327		
-328	-6	5	-331	3	-2	-335	4	-2
-339			-340	-4	2	-344		
-347	-2	-2	-355			-356		
-359			-367			-371	2	0
-372	-1	3	-376	4	-3	-379		
-383			-388			-391	-	-
-395	2	0	-399	-6	4	-403	1	2
-404	-2	0	-407	-2	-2	-408		
-411			-415	2	-2	-419		
-420	2	0	-424			-427	-4	4
-431			-435			-436		
-439	1	0	-440	-2	2	-443	5	4
-447			-451			-452		
-455	0	2	-456			-463	-6	2
-467			-471			-472	-4	0
-479			-483	-	-	-487	-1	0
-488			-491	-1	-2	-499	3	-2

$X_0(23)$: Even quadratic characters

$$\Lambda(\chi) = (3+\omega)\cdot(a+b\omega)^2 \; ; \; \omega = (1+\sqrt{5})/2$$

m_χ	$a + b\cdot(\omega)$		m_χ	$a + b\cdot(\omega)$		m_χ	$a + b\cdot(\omega)$	
5			8	2	-1	12	1	-1
13	1	0	17			21		
24	2	-1	28			29	3	-2
33			37			40		
41	3	-2	44			53		
56			57			60		
61			65			69	-	-
73	1	0	76			77	2	-2
85	2	-2	88			89		
92	-	-	93	1	0	97		
101	4	-2	104	2	-1	105	4	-2
109			113			120		
124	1	-1	129			133	2	0
136			137			140	6	-4
141	1	0	145			149		
152			156	5	-3	157		
161	-	-	165	4	-2	168		
172			173	2	-2	177	0	0
181			184	-	-	185	4	-2
188	1	-1	193	3	-2	197	3	-2
201			204			205		
209	0	0	213	1	-2	217		
220	2	0	221			229		
232	0	1	233	7	-4	236	0	0
237			241			248	4	-3
249			253	-	-	257	1	0
264			265	0	0	268		
269	1	0	273			277	1	-2
280	2	-2	281			284	5	-3
285	2	0	293			296		
301	0	0	305	6	-4	309		
312	2	-1	313			316		
317	4	-2	321			328	0	1
329			332			337		
341			344			345	-	-
348	3	-3	349	1	0	353	1	0
357	2	-2	364			365		
373			376	4	-3	377	3	-2
380	10	-6	381	1	0	385		
389			393	5	-4	397	1	0
401			408			409	1	0
412			413			417	7	-4
421			424			428		
429			433			437	-	-
440	2	-2	444			445	2	0
449	0	0	453	3	-2	456		
457			460	-	-	461	1	0
465			469	4	-2	472	0	2
473	6	-4	476	4	-2	481		
485	4	-2	488			489	3	-2
492	3	-1	493			497		

$X_0(29)$: Odd quadratic characters

$$\Lambda(\chi) = (a+b\sqrt{2})^2$$

m_χ	a +	$b\cdot(\sqrt{2})$	m_χ	a +	$b\cdot(\sqrt{2})$	m_χ	a +	$b\cdot(\sqrt{2})$
-3	1	0	-4			-7		
-8	0	1	-11	-1	-1	-15	-1	0
-19	0	1	-20			-23		
-24			-31	-1	1	-35		
-39	1	-1	-40	2	-1	-43	-1	-1
-47	1	0	-51			-52		
-55	1	-1	-56	-2	0	-59		
-67			-68	-2	0	-71		
-79	1	0	-83			-84	-2	0
-87	-	-	-88			-91		
-95	0	1	-103			-104	0	-1
-107			-111			-115		
-116			-119	2	0	-120		
-123			-127	-2	1	-131	-2	1
-132			-136			-139		
-143	1	-2	-148	0	2	-151		
-152			-155	3	3	-159	-1	1
-163	3	0	-164	2	-2	-167		
-168			-179			-183		
-184	-2	0	-187			-191	0	-1
-195	1	-1	-199			-203	-	-
-211	1	-2	-212			-215	3	-1
-219			-223			-227		
-228			-231			-232	-	-
-235	-1	0	-239			-244	-2	2
-247	0	-1	-248			-251	1	2
-255			-259	-2	0	-260		
-263	1	1	-264	0	1	-267		
-271	1	-1	-276	2	-2	-280	-2	0
-283			-287	0	0	-291		
-292	4	-2	-295			-296		
-299			-303			-307	-1	1
-308	2	-2	-311	-2	1	-312		
-319	-	-	-323			-327	1	0
-328			-331	3	2	-335		
-339			-340	-2	0	-344		
-347			-355			-356	2	2
-359	1	0	-367	2	-1	-371		
-372			-376			-379	-2	-1
-383			-388	-2	0	-391	0	0
-395	-5	-2	-399			-403	-1	0
-404	0	0	-407			-408	-2	0
-411			-415			-419		
-420	-2	4	-424	0	-1	-427	0	2
-431			-435	-	-	-436		
-439			-440			-443	-2	1
-447	3	-1	-451			-452	-4	0
-455			-456	2	-2	-463		
-467	3	-1	-471			-472	-2	2
-479	1	-1	-483	0	4	-487		
-488			-491	3	1	-499		

$\chi_0(29)$: Even quadratic characters

$$\Lambda(\chi) = (3-\sqrt{2})\cdot(a+b\sqrt{2})^2$$

m_χ	a + b·($\sqrt{2}$)		m_χ	a = b·($\sqrt{2}$)		m_χ	a + b·($\sqrt{2}$)	
5	1	0	8			12		
13	1	1	17			21		
24	0	1	28	2	0	29	-	-
33	1	0	37			40		
41			44			53	1	-1
56			57	2	-1	60		
61			65	1	-1	69		
73			76			77		
85			88	2	-1	89		
92	2	-2	93	1	2	97		
101			104			105		
109	1	2	113			120	2	-1
124			129	1	0	133		
136	2	0	137			140	2	0
141	1	-1	145	-	-	149	1	1
152	2	0	156			157		
161	0	0	165	3	0	168	2	-2
172			173	0	2	177		
181	1	2	184			185		
188			193			197	0	2
201			204	2	-2	205		
209	2	-1	213			217		
220			221			229		
232	-	-	233	1	0	236	2	0
237	1	3	241	1	0	248	0	1
249			253			257	1	-1
264			265	1	1	268	0	2
269			273			277	4	0
280			281	1	0	284	0	0
285	2	1	293			296	4	-2
301			305			309		
312	4	-1	313	3	-1	316		
317			321			328	2	2
329			332	2	-2	337		
341	1	-1	344	2	-1	345		
348	-	-	349	5	3	353	2	0
357	2	2	364	2	-2	365		
373	1	0	376	2	1	377	-	-
380			381	4	1	385		
389			393	2	-1	397	5	4
401	3	-2	408			409		
412	0	2	413	2	0	417		
421			424			428	2	-2
429	3	-1	433			437		
440	0	1	444	0	2	445		
449			453			456		
457	2	0	460	2	2	461		
465	3	-2	469	2	0	472		
473	3	-1	476			481		
485			488	2	-2	489	1	-1
492	2	0	493	-	-	497	2	-2

$\chi_0(31)$: Odd quadratic characters

$\Lambda(\chi) = (a+b\omega)^2$; $\omega = (1+\sqrt{5})/2$

m_χ	a +	b·(ω)	m_χ	a +	b·(ω)	m_χ	a +	b·(ω)
-3			-4	-1	0	-7	0	1
-8	1	-1	-11			-15		
-19	0	1	-20	1	0	-23		
-24			-31	-	-	-35	-2	1
-39	2	0	-40	-1	-1	-43		
-47	0	0	-51	2	-2	-52		
-55			-56	-1	1	-59	2	-1
-67	-2	0	-68			-71	-2	1
-79			-83			-84		
-87	-2	0	-88			-91		
-95	2	-1	-103	2	1	-104		
-107	2	-1	-111	0	-2	-115		
-116			-119			-120		
-123			-127			-131	0	0
-132	2	0	-136			-139		
-143	0	0	-148			-151		
-152	1	-1	-155	-	-	-159	0	0
-163	0	1	-164	-1	0	-167		
-168			-179			-183	-2	2
-184			-187	0	2	-191	2	-1
-195	2	0	-199			-203		
-211	2	-1	-212			-215		
-219	-4	2	-223			-227	0	0
-228			-231	0	2	-232		
-235	0	2	-239			-244		
-247			-248	-	-	-251		
-255	2	-2	-259			-260		
-263			-264	-2	2	-267	2	-2
-271			-276	-2	2	-280	1	-3
-283	0	-2	-287	0	-1	-291		
-292			-295	0	3	-296		
-299	-2	2	-303			-307	2	-1
-308			-311	0	-1	-312	-4	2
-319	4	2	-323			-327		
-328	-1	1	-331			-335	-2	2
-339			-340			-344		
-347			-355	0	-1	-356		
-359	-2	3	-367			-371		
-372	-	-	-376	4	0	-379	0	-2
-383			-388	1	2	-391	-2	-2
-395			-399			-403	-	-
-404	3	-2	-407	4	-2	-408	2	0
-411	-2	0	-415			-419	0	-1
-420			-424			-427		
-431	-2	0	-435	2	0	-436	1	4
-439	2	1	-440			-443	4	-3
-447			-451			-452	-1	0
-455			-456			-463		
-467	-2	1	-471			-472	-1	3
-479	2	1	-483	2	0	-487		
-488			-491			-499		

$\chi_0(31)$: Even quadratic characters

$\Lambda(\chi) = (1-2\omega)\cdot(a+b\omega)^2$; $\omega = (1+\sqrt{5})/2$

m_χ	a +	b·(ω)	m_χ	a +	b·(ω)	m_χ	a +	b·(ω)
5	1	0	8	0	1	12		
13			17			21		
24			28	1	-1	29		
33	2	0	37			40	2	-1
41	1	2	44			53		
56	2	1	57			60		
61			65			69	2	-2
73			76	3	-1	77		
85			88			89		
92			93	-	-	97	1	0
101	1	-2	104			105		
109	1	0	113	1	2	120		
124	-	-	129	2	-2	133	1	0
136			137			140	3	1
141			145			149	0	2
152	2	-3	156	0	2	157	3	-2
161			165	2	0	168		
172			173	0	2	177		
181			184			185		
188	2	2	193	1	0	197		
201			204	2	0	205	1	0
209			213			217	-	-
220			221	0	2	229		
232			233	1	2	236	1	-5
237	4	-2	241			248	-	-
249	0	0	253	0	0	257	1	4
264	4	-2	265			268	2	0
269			273	2	-2	277		
280	0	1	281	3	2	284	3	3
285			293	2	-2	296		
301			305			309		
312	2	0	313			316		
317	1	2	321			328	4	-1
329	2	0	332			337		
341	-	-	344			345	2	2
348	4	-2	349	4	-2	353		
357	0	2	364			365		
373	3	-2	376	4	-2	377	4	2
380	1	1	381	2	0	385		
389			393			397	3	-2
401			408	2	-2	409		
412	5	-3	413	1	-2	417	2	0
421	3	-2	424			428	1	3
429			433			437		
440			444	6	-2	445		
449			453	2	-2	456		
457			460			461		
465	-	-	469	2	-2	472	0	1
473	2	4	476			481	2	0
485	1	2	488			489		
492			493	4	-2	497	1	2

$X_0(37)$: Odd quadratic characters

$$\Lambda(\chi) = (a^2, 2b^2) \in \mathbb{Z} \times \mathbb{Z}$$

m_χ	a	b	m_χ	a	b	m_χ	a	b
-3	1		-4	1		-7	1	
-8		-1	-11	1		-15		1
-19		-1	-20		1	-23		0
-24		1	-31		0	-35		1
-39		-1	-40	2		-43		-1
-47	1		-51		1	-52		1
-55		-1	-56		-1	-59		0
-67	6		-68		-1	-71	1	
-79		1	-83	1		-84	1	
-87		0	-88		1	-91		1
-95	0		-103		1	-104	0	
-107	0		-111	-	-	-115	6	
-116		0	-119		-1	-120	2	
-123	3		-127	1		-131		1
-132	3		-136	4		-139	0	
-143		-1	-148	-	-	-151	2	
-152	2		-155	2		-159	1	
-163		-1	-164	1		-167		1
-168		-1	-179		-2	-183		-2
-184	0		-187		1	-191		2
-195	2		-199		0	-203		2
-211	3		-212	3		-215	0	
-219	1		-223	3		-227		-2
-228		-1	-231	1		-232	6	
-235		1	-239		0	-244		0
-247	4		-248	0		-251		-1
-255	0		-259	-	-	-260	2	
-263	1		-264		1	-267		1
-271	1		-276		-2	-280	6	
-283		0	-287	1		-291		2
-292	1		-295	4		-296	-	-
-299	2		-303	1		-307	5	
-308	1		-311		-1	-312	4	
-319		2	-323	2		-327		3
-328		-1	-331		0	-335		0
-339		0	-340	4		-344	0	
-347		-2	-355		-1	-356		3
-359	1		-367	2		-371	1	
-372		2	-376		1	-379	7	
-383		-2	-388		2	-391	0	
-395	2		-399		-1	-403	6	
-404	1		-407	-	-	-408	4	
-411	2		-415		-1	-419	3	
-420		1	-424		3	-427		-2
-431		3	-435	2		-436		3
-439		0	-440	2		-443	1	
-447	1		-451	5		-452		-2
-455	2		-456	2		-463		2
-467		2	-471	1		-472	2	
-479		2	-483		2	-487		-1
-488	2		-491	2		-499		0

$X_0(37)$: Even quadratic characters

$\Lambda(\chi) = (2a^2, 3b^2) \in \mathbb{Z} \times \mathbb{Z}$

m_χ	a	b	m_χ	a	b	m_χ	a	b
5	1		8	1		12		1
13	1		17	1		21		1
24	1		28		1	29	2	
33		1	37	-	-	40		0
41		1	44		1	53		1
56	1		57	1		60	1	
61	0		65		0	69	0	
73		1	76	1		77		1
85		2	88	1		89	1	
92	2		93	2		97	0	
101		1	104		2	105	1	
109	1		113	0		120		0
124	0		129	1		133	1	
136		2	137		2	140	3	
141		1	145		2	149		1
152		0	156	1		157		1
161	2		165	1		168	1	
172	1		173		1	177	0	
181		1	184		0	185	-	-
188		3	193	1		197		1
201		0	204	1		205	1	
209	1		213		3	217	0	
220	1		221		0	229		3
232		2	233		0	236	2	
237	3		241	0		248		0
249		1	253	0		257	2	
264	1		265	1		268		2
269		0	273	1		277	2	
280		4	281	2		284		1
285		2	293		2	296	-	-
301	1		305		2	309	1	
312		2	313	1		316	1	
317		2	321		0	328	1	
329		1	332		1	337		1
341	4		344		2	345		2
348	2		349		0	353	1	
357	3		364	1		365	3	
373		1	376	1		377		0
380		2	381		1	385	1	
389	2		393	1		397		3
401	1		408		2	409	0	
412	1		413	6		417		0
421	1		424	1		428		0
429	3		433		1	437		2
440		4	444	-	-	445		0
449	1		453		0	456		0
457	1		460		2	461	5	
465		2	469		0	472		2
473	3		476	1		481	-	-
485		2	488		2	489	1	
492		3	493		0	497		1

$X_0(41)$: Odd quadratic characters

$$\Lambda(\chi) = (a+b\omega+c\omega^2)^2$$

$$(\omega^3-4\omega+2 = 0)$$

m_χ	a +	$b\omega$ +	$c\omega^2$	m_χ	a +	$b\omega$ +	$c\omega^2$	m_χ	a +	$b\omega$ +	$c\omega^2$
-3	1	0	0	-4				-7	-1	1	0
-8				-11	3	-1	-1	-15	0	-1	0
-19	-3	0	1	-20				-23			
-24	2	0	-1	-31				-35	-4	1	1
-39				-40				-43			
-47	1	0	-1	-51				-52	-2	2	0
-55	0	-1	1	-56	-2	2	1	-59			
-67	3	-1	-1	-68	0	-2	0	-71	1	0	0
-79	1	0	-1	-83				-84			
-87				-88	0	0	1	-91			
-95	0	1	0	-103				-104	-2	0	0
-107				-111	0	1	0	-115			
-116	-2	0	0	-119				-120	2	0	0
-123	-	-	-	-127				-131			
-132				-136	0	-2	0	-139			
-143				-148				-151	1	0	0
-152	2	-2	-1	-155				-159			
-163				-164	-	-	-	-167	1	-1	1
-168				-179	1	1	0	-183	0	0	-1
-184				-187				-191	-1	1	1
-195				-199	-1	1	-1	-203			
-211	-7	1	2	-212	-2	2	2	-215			
-219	8	-1	-2	-223				-227	7	0	-2
-228				-231	0	-1	0	-232	-2	2	0
-235	2	1	0	-239	1	0	-1	-244			
-247				-248				-251			
-255				-259	4	1	-1	-260	4	-2	-2
-263	-1	0	0	-264				-267			
-271				-276	4	-2	-2	-280	2	0	0
-283				-287	-	-	-	-291			
-292				-295				-296			
-299	-8	0	2	-303				-307			
-308				-311	1	-1	0	-312			
-319				-323				-327			
-328	-	-	-	-331	5	-1	-2	-335	2	-3	-1
-339	-6	3	2	-340	4	0	-2	-344			
-347	1	0	-1	-355	6	-3	-2	-356	-4	2	2
-359				-367				-371			
-372	0	-2	0	-376	0	0	-1	-379			
-383	-1	3	0	-388	0	0	2	-391	2	0	0
-395	-12	3	4	-399	-2	1	1	-403	-6	0	2
-404	2	0	0	-407	-2	1	1	-408			
-411				-415				-419			
-420				-424	-2	0	0	-427	0	0	1
-431				-435				-436	-2	0	0
-439	1	-3	1	-440	-2	2	0	-443			
-447				-451	-	-	-	-452			
-455				-456				-463	1	0	0
-467				-471				-472			
-479	1	1	0	-483				-487			
-488				-491				-499	9	-2	-2

$\chi_0(41)$: Even quadratic characters

$$\Lambda(\chi) = (4-7\omega-\omega^2)\cdot(a+b\omega+c\omega^2)^2$$

$$(\omega^3-4\omega+2 = 0)$$

m_χ	a +	$b\omega$ +	$c\omega^2$	m_χ	a +	$b\omega$ +	$c\omega^2$	m_χ	a +	$b\omega$ +	$c\omega^2$
5	7	-1	-2	8	4	-1	-1	12			
13				17				21	15	-2	-4
24				28				29			
33	3	0	-1	37	19	-3	-5	40	4	0	-1
41	-	-	-	44				53			
56				57	1	-1	0	60			
61	18	-3	-5	65				69			
73	3	-1	-1	76				77	11	-1	-3
85				88				89			
92	6	0	-2	93				97			
101				104				105	6	-1	-2
109				113	1	1	-1	120			
124	14	-2	-4	129				133	19	-2	-5
136				137				140			
141	37	-5	-10	145				149			
152				156	2	0	0	157			
161				165	30	-5	-8	168	2	0	-1
172	2	0	0	173	14	-2	-4	177			
181				184	14	-2	-4	185	4	-1	-1
188				193				197	20	-3	-5
201	11	-2	-3	204	8	-2	-2	205	-	-	-
209	1	0	0	213	33	-5	-9	217			
220				221	38	-6	-10	229			
232				233				236	14	-2	-4
237	9	-1	-2	241	4	-1	-1	248	2	0	0
249				253				257			
264	4	0	-1	265				268			
269	34	-5	-9	273				277	1	1	0
280				281				284			
285	12	-1	-3	293				296	12	-2	-3
301				305	2	0	-1	309			
312	10	-2	-2	313				316			
317				321				328	-	-	-
329	5	-2	-1	332	0	0	0	337	7	-1	-2
341				344	6	0	-2	345			
348	6	-2	-2	349	23	-3	-6	353	1	-1	0
357				364	14	-2	-4	365	20	-3	-5
373	14	-2	-4	376				377	6	0	-2
380				381				385	10	-1	-3
389	30	-4	-8	393				397			
401	13	-1	-4	408	20	-2	-6	409	3	-1	-1
412	8	-2	-2	413				417			
421				424				428	4	0	-2
429				433	8	-2	-2	437			
440				444				445			
449	2	1	-1	453	27	-3	-7	456	18	-2	-5
457				460	16	-2	-4	461	3	-1	-1
465				469	17	-3	-5	472	6	-2	-2
473				476	16	-2	-4	481			
485				488	14	-2	-4	489			
492	-	-	-	493	30	-4	-8	497	5	0	-2

$X_0(43)$: Odd quadratic characters

$$\Lambda(\chi) = (2a^2, \sqrt{2}\cdot(b+c\sqrt{2})^2)$$

m_χ	a	b + $c\sqrt{2}$		m_χ	a	b + $c\sqrt{2}$		m_χ	a	b + $c\sqrt{2}$	
-3	1			-4		1	0	-7	1		
-8	1			-11		-1	-1	-15		0	1
-19	2			-20	1			-23		-1	0
-24		0	1	-31		-1	0	-35		0	1
-39	1			-40		0	1	-43	-	-	-
-47		0	-1	-51	1			-52		1	-1
-55	1			-56		-2	-1	-59		2	1
-67		-1	-1	-68		1	0	-71	0		
-79		0	-1	-83		-1	0	-84		-2	-1
-87		2	-1	-88	3			-91	1		
-95		2	0	-103		-1	1	-104	1		
-107		2	1	-111		2	0	-115	3		
-116	1			-119	1			-120	0		
-123	5			-127		-1	1	-131	0		
-132	3			-136	1			-139		-1	0
-143		-1	0	-148	4			-151	1		
-152		-2	0	-155	1			-159	1		
-163	7			-164		-1	-1	-167		1	0
-168	2			-179	1			-183		-2	1
-184	3			-187		1	-1	-191	0		
-195		-2	-1	-199	0			-203		-2	-1
-211	2			-212		-1	-2	-215	-	-	-
-219		4	1	-223	3			-227	4		
-228		0	2	-231		0	-1	-232		0	1
-235	0			-239		-2	1	-244	1		
-247	2			-248	3			-251		3	1
-255		0	-1	-259		0	2	-260	1		
-263	2			-264		0	-3	-267		0	1
-271		-3	1	-276	1			-280	2		
-283		-1	0	-287	1			-291	3		
-292	7			-295	2			-296		0	-2
-299		-1	-1	-303	1			-307		-1	0
-308	1			-311		-1	1	-312		-2	-1
-319	1			-323	4			-327	3		
-328	3			-331	3			-335	1		
-339		-2	0	-340	5			-344	-	-	-
-347	3			-355		0	2	-356	1		
-359		-3	-2	-367		0	1	-371	3		
-372	3			-376	2			-379		-3	-3
-383	0			-388		1	0	-391		1	-2
-395	2			-399	0			-403		1	-1
-404		1	-2	-407	0			-408		4	1
-411		4	2	-415	1			-419	1		
-420	4			-424	1			-427		0	1
-431		3	1	-435	6			-436		1	1
-439		3	-2	-440		0	-1	-443		0	-1
-447		0	0	-451		-1	-2	-452	0		
-455		2	-1	-456	2			-463	3		
-467	4			-471		-2	-2	-472	6		
-479		1	0	-483		-2	-1	-487		0	0
-488		2	1	-491	1			-499	3		

$\chi_0(43)$: Even quadratic characters

$$\Lambda(\chi) = (2a^2, (4-\sqrt{2})\cdot(b+c\sqrt{2})^2)$$

m_χ	a	$b + c\sqrt{2}$		m_χ	a	$b + c\sqrt{2}$		m_χ	a	$b + c\sqrt{2}$	
5	1			8	1			12	1		
13		1	1	17		1	0	21		2	1
24		2	1	28	1			29	1		
33	1			37	2			40		2	1
41		1	1	44		1	1	53		1	0
56		0	1	57		0	0	60		0	1
61	1			65	1			69	1		
73	1			76	0			77	3		
85	1			88	1			89	1		
92		1	0	93	3			97		1	0
101		3	2	104	1			105	0		
109		1	1	113	2			120	0		
124		3	2	129	-	-	-	133		0	0
136	1			137	2			140		0	1
141	2			145		2	1	149	0		
152		2	0	156	1			157	2		
161	1			165		2	1	168	2		
172	-	-	-	173		0	0	177	2		
181		2	1	184	1			185		2	0
188		2	1	193		1	0	197		2	1
201	1			204	1			205	1		
209	0			213		2	2	217	1		
220	1			221		1	1	229		5	3
232		2	1	233	0			236		4	1
237	0			241	0			248	1		
249	1			253		1	1	257	1		
264		2	3	265	1			268		1	1
269		3	3	273		0	1	277	1		
280	2			281		1	-1	284	2		
285	2			293		0	1	296		0	0
301	-	-	-	305		2	1	309	1		
312		0	1	313	1			316		2	1
317		1	1	321	0			328	1		
329	0			332		1	2	337		1	1
341		1	-1	344	-	-	-	345		0	1
348		2	1	349	1			353		1	-1
357		2	1	364	1			365		0	1
373	2			376	0			377	3		
380		2	2	381	1			385		2	1
389	0			393		0	2	397		2	1
401		1	-1	408		2	1	409	1		
412		1	-1	413	4			417	1		
421	2			424	1			428		0	1
429	1			433	1			437	6		
440		2	3	444		2	2	445		2	1
449	1			453		4	3	456	2		
457	1			460	1			461		4	3
465		0	1	469	1			472	0		
473	-	-	-	476	3			481	0		
485	1			488		0	1	489		4	1
492	1			493	3			497		2	0

§6.2. Genus One Curves, $X_0(N)$

$N = 14, 15, 20, 21, 24, 27, 32, 36, 49$

The special values

$$\Lambda(\chi) = \frac{\tau(\bar{\chi})\, L(f, \chi, 1)}{\Omega_f^{\operatorname{sgn} \chi}}$$

are tabulated for quadratic characters of conductor m_χ with $(m_\chi, N) = 1$ and $|m_\chi| < 500$.

The period $\Omega_f^{\operatorname{sgn}(\chi)}$ corresponds to a generator of $H_1(X_0(N), \mathbb{Z})^{\operatorname{sgn}(\chi)}$ such that $\Lambda(\chi) \geq 0$ for all χ.

When $N = 36$ or 49 and χ is an odd quadratic character of conductor prime to N, the sign of the functional equation for $L(f, \chi, s)$ implies $\Lambda(\chi) = 0$. Hence tables of special values have not been included for these cases.

The curves $X_0(11)$, $X_0(17)$, $X_0(19)$ also have genus one: the special values of their L-functions have been tabulated in §6.1.

$\Gamma_0(14)$: Special values for

Odd quadratic characters

m_χ	$\Lambda(\chi)$	m_χ	$\Lambda(\chi)$	m_χ	$\Lambda(\chi)$	m_χ	$\Lambda(\chi)$	m_χ	$\Lambda(\chi)$
-3	1	-11	0	-15	2	-19	9	-23	0
-31	0	-39	2	-43	0	-47	0	-51	0
-55	0	-59	9	-67	0	-71	2	-79	2
-83	9	-87	0	-95	2	-103	0	-107	0
-111	0	-115	36	-123	0	-127	8	-131	9
-139	9	-143	0	-151	8	-155	0	-159	0
-163	0	-167	0	-179	0	-183	2	-187	36
-191	2	-195	36	-199	0	-211	0	-215	0
-219	0	-223	0	-227	9	-235	0	-239	0
-247	18	-251	9	-255	0	-263	2	-267	0
-271	0	-283	9	-291	0	-295	2	-299	0
-303	2	-307	81	-311	0	-319	32	-323	0
-327	0	-331	0	-335	0	-339	0	-347	0
-355	36	-359	8	-367	0	-379	0	-383	0
-391	0	-395	36	-403	0	-407	8	-411	36
-415	2	-419	9	-431	0	-435	0	-439	0
-443	0	-447	0	-451	36	-463	2	-467	9
-471	2	-479	0	-487	8	-491	0	-499	0

$\Gamma_0(14)$: Special Values for

Even quadratic characters

m_χ	$\Lambda(\chi)$	m_χ	$\Lambda(\chi)$	m_χ	$\Lambda(\chi)$	m_χ	$\Lambda(\chi)$	m_χ	$\Lambda(\chi)$
5	3	13	3	17	0	29	0	33	0
37	0	41	0	53	0	57	6	61	3
65	6	69	12	73	0	85	0	89	0
93	0	97	0	101	27	109	0	113	6
129	0	137	24	141	0	145	0	149	0
157	3	165	0	173	3	177	6	181	3
185	0	193	6	197	0	201	0	205	0
209	0	213	48	221	0	229	27	233	0
237	12	241	0	249	6	253	0	257	0
265	0	269	27	277	0	281	24	285	12
293	75	305	54	309	0	313	0	317	0
321	0	337	6	341	12	345	0	349	27
353	0	365	0	373	0	377	0	381	0
389	0	393	6	397	3	401	6	409	0
417	6	421	0	429	0	433	0	437	48
445	0	449	24	453	48	457	6	461	147
465	0	473	0	481	0	485	0	489	0
493	12								

$\Gamma_0(15)$: Special values for

Odd quadratic characters

m_χ	$\Lambda(\chi)$	m_χ	$\Lambda(\chi)$	m_χ	$\Lambda(\chi)$	m_χ	$\Lambda(\chi)$	m_χ	$\Lambda(\chi)$
-4	2	-7	0	-8	2	-11	0	-19	8
-23	2	-31	8	-43	0	-47	2	-52	0
-56	0	-59	0	-67	0	-68	0	-71	0
-79	8	-83	2	-88	0	-91	32	-103	0
-104	0	-107	2	-116	0	-119	0	-127	0
-131	0	-136	32	-139	8	-143	8	-148	0
-151	8	-152	8	-163	0	-164	0	-167	2
-179	0	-184	32	-187	0	-191	0	-199	8
-203	32	-211	8	-212	8	-223	0	-227	2
-232	0	-239	0	-244	32	-247	0	-248	0
-251	0	-259	0	-263	2	-271	8	-283	0
-287	8	-292	0	-296	0	-299	0	-307	0
-308	0	-311	0	-319	32	-323	32	-328	0
-331	72	-344	0	-347	2	-356	0	-359	0
-367	0	-371	0	-376	32	-379	72	-383	18
-388	0	-391	0	-403	0	-404	0	-407	0
-419	0	-424	0	-427	0	-431	0	-436	32
-439	8	-443	2	-451	32	-452	32	-463	0
-467	18	-472	0	-479	0	-487	0	-488	8
-491	0	-499	72						

$\Gamma_0(15)$: Special values for

Even quadratic characters

m_χ	$\Lambda(\chi)$	m_χ	$\Lambda(\chi)$	m_χ	$\Lambda(\chi)$	m_χ	$\Lambda(\chi)$	m_χ	$\Lambda(\chi)$
8	4	13	0	17	4	28	0	29	0
37	0	41	0	44	0	53	4	56	0
61	16	73	0	76	16	77	16	88	0
89	0	92	16	97	0	101	0	104	0
109	16	113	4	124	16	133	0	136	0
137	4	149	0	152	0	157	0	161	0
172	0	173	36	181	0	184	16	188	16
193	0	197	36	209	0	217	0	221	0
229	0	232	0	233	4	236	0	241	16
248	16	253	0	257	4	268	0	269	0
277	0	281	0	284	0	293	36	296	0
301	16	313	0	316	16	317	4	328	0
329	0	332	64	337	0	341	0	344	0
349	64	353	36	364	0	373	0	376	16
377	0	389	0	397	0	401	0	409	16
412	0	413	0	421	16	424	64	428	0
433	0	437	16	449	0	457	0	461	0
469	16	472	0	473	16	476	0	481	0
488	16	493	0	497	16				

$\Gamma_0(20)$: Special values for

Odd quadratic characters

m_χ	$\Lambda(\chi)$	m_χ	$\Lambda(\chi)$	m_χ	$\Lambda(\chi)$	m_χ	$\Lambda(\chi)$	m_χ	$\Lambda(\chi)$
-3	1	-7	3	-11	0	-19	0	-23	3
-31	0	-39	0	-43	9	-47	3	-51	0
-59	0	-67	9	-71	0	-79	0	-83	9
-87	12	-91	0	-103	27	-107	9	-111	0
-119	0	-123	0	-127	3	-131	0	-139	0
-143	12	-151	0	-159	0	-163	9	-167	3
-179	0	-183	12	-187	36	-191	0	-199	0
-203	0	-211	0	-219	0	-223	3	-227	9
-231	0	-239	0	-247	12	-251	0	-259	0
-263	3	-267	36	-271	0	-283	81	-287	0
-291	0	-299	0	-303	12	-307	9	-311	0
-319	0	-323	0	-327	12	-331	0	-339	0
-347	9	-359	0	-367	75	-371	0	-379	0
-383	3	-391	0	-399	0	-403	36	-407	0
-411	0	-419	0	-427	0	-431	0	-439	0
-443	9	-447	48	-451	0	-463	3	-467	9
-471	0	-479	0	-483	36	-487	27	-491	0
-499	0								

$\Gamma_0(20)$: Special values for

Even quadratic characters

m_χ	$\Lambda(\chi)$	m_χ	$\Lambda(\chi)$	m_χ	$\Lambda(\chi)$	m_χ	$\Lambda(\chi)$	m_χ	$\Lambda(\chi)$
13	0	17	0	21	2	29	8	33	0
37	0	41	6	53	0	57	0	61	2
69	2	73	0	77	0	89	0	93	0
97	0	101	0	109	2	113	0	129	6
133	0	137	0	141	18	149	2	157	0
161	6	173	0	177	0	181	8	193	0
197	0	201	6	209	24	213	0	217	0
221	8	229	0	233	0	237	0	241	6
249	6	253	0	257	0	269	18	273	0
277	0	281	6	293	0	301	2	309	2
313	0	317	0	321	6	329	6	337	0
341	8	349	0	353	0	357	0	373	0
377	0	381	18	389	18	393	0	397	0
401	24	409	6	413	0	417	0	421	2
429	8	433	0	437	0	449	54	453	0
457	0	461	8	469	18	473	0	481	0
489	6	493	0	497	0				

$\Gamma_0(21)$: Special values for

Odd quadratic characters

m_χ	$\Lambda(\chi)$	m_χ	$\Lambda(\chi)$	m_χ	$\Lambda(\chi)$	m_χ	$\Lambda(\chi)$	m_χ	$\Lambda(\chi)$
-4	0	-8	4	-11	4	-19	2	-20	0
-23	4	-31	2	-40	8	-43	0	-47	0
-52	8	-55	0	-59	0	-67	0	-68	0
-71	4	-79	0	-83	0	-88	0	-95	16
-103	2	-104	0	-107	36	-115	8	-116	16
-127	0	-131	0	-136	0	-139	2	-143	0
-148	0	-151	0	-152	0	-155	16	-163	0
-164	0	-167	0	-179	4	-184	0	-187	8
-191	4	-199	2	-211	0	-212	16	-215	0
-223	2	-227	0	-232	0	-235	0	-239	4
-244	8	-247	0	-248	0	-251	0	-260	64
-263	4	-271	2	-283	18	-292	32	-295	0
-296	16	-299	0	-307	18	-311	0	-319	0
-323	16	-328	0	-331	0	-335	0	-340	0
-344	16	-347	36	-355	0	-356	0	-359	4
-367	2	-376	0	-379	0	-383	0	-388	32
-391	8	-395	0	-403	0	-404	0	-407	0
-415	0	-419	0	-424	0	-431	36	-436	0
-439	2	-440	0	-443	4	-451	8	-452	64
-463	0	-467	0	-472	8	-479	0	-487	0
-488	0	-491	36	-499	0				

$\Gamma_0(21)$: Special values for

Even quadratic characters

m_χ	$\Lambda(\chi)$	m_χ	$\Lambda(\chi)$	m_χ	$\Lambda(\chi)$	m_χ	$\Lambda(\chi)$	m_χ	$\Lambda(\chi)$
5	2	8	0	13	0	17	2	29	0
37	16	40	0	41	2	44	0	53	0
61	0	65	0	73	0	76	0	85	16
88	16	89	2	92	0	97	0	101	2
104	8	109	0	113	0	124	0	136	0
137	0	145	0	149	0	152	8	157	0
172	16	173	2	181	0	184	16	185	8
188	32	193	16	197	0	205	16	209	0
220	0	221	0	229	0	232	64	233	0
236	8	241	0	248	32	253	16	257	2
265	0	268	16	269	18	277	16	281	0
284	0	293	18	296	0	305	0	313	0
316	16	317	0	328	0	332	8	337	0
341	8	344	0	349	0	353	2	365	0
373	0	376	0	377	32	380	0	389	0
397	0	401	0	409	0	412	0	421	16
424	0	428	0	433	0	437	32	440	0
445	16	449	0	457	0	460	0	461	2
472	0	473	0	481	0	485	0	488	72
493	0								

$\Gamma_0(24)$: Special values for

Odd quadratic characters

m_χ	$\Lambda(\chi)$	m_χ	$\Lambda(\chi)$	m_χ	$\Lambda(\chi)$	m_χ	$\Lambda(\chi)$	m_χ	$\Lambda(\chi)$
-7	4	-11	2	-19	0	-23	0	-31	4
-35	8	-43	0	-47	0	-55	16	-59	2
-67	0	-71	0	-79	4	-83	18	-91	0
-95	0	-103	4	-107	2	-115	0	-119	0
-127	36	-131	2	-139	0	-143	0	-151	36
-155	0	-163	0	-167	0	-179	2	-187	0
-191	0	-199	4	-203	0	-211	0	-215	0
-223	36	-227	18	-235	0	-239	0	-247	16
-251	2	-259	0	-263	0	-271	36	-283	0
-287	0	-295	16	-299	0	-307	0	-311	0
-319	0	-323	32	-331	0	-335	0	-347	50
-355	0	-359	0	-367	4	-371	32	-379	0
-383	0	-391	16	-395	32	-403	0	-407	0
-415	64	-419	2	-427	0	-431	0	-439	36
-443	18	-451	0	-455	0	-463	100	-467	2
-479	0	-487	36	-491	18	-499	0		

$\Gamma_0(24)$: Special values for

Even quadratic characters

m_χ	$\Lambda(\chi)$	m_χ	$\Lambda(\chi)$	m_χ	$\Lambda(\chi)$	m_χ	$\Lambda(\chi)$	m_χ	$\Lambda(\chi)$
5	2	13	0	17	0	29	2	37	0
41	0	53	18	61	0	65	0	73	0
77	8	85	0	89	0	97	16	101	2
109	0	113	0	133	0	137	0	145	16
149	18	157	0	161	0	173	18	181	0
185	0	193	0	197	2	205	0	209	0
217	16	221	32	229	0	233	0	241	0
253	0	257	0	265	16	269	2	277	0
281	0	293	50	301	0	305	0	313	16
317	50	329	0	337	16	341	32	349	0
353	0	365	8	373	0	377	0	385	0
389	2	397	0	401	0	409	0	413	32
421	0	433	16	437	8	445	0	449	0
457	0	461	2	469	0	473	0	481	0
485	8	493	0	497	0				

$\Gamma_0(27)$: Special values for

Odd quadratic characters

m_χ	$\Lambda(\chi)$	m_χ	$\Lambda(\chi)$	m_χ	$\Lambda(\chi)$	m_χ	$\Lambda(\chi)$	m_χ	$\Lambda(\chi)$
-4	1	-7	1	-8	0	-11	0	-19	1
-20	0	-23	0	-31	0	-35	0	-40	4
-43	4	-47	0	-52	1	-55	4	-56	0
-59	0	-67	1	-68	0	-71	0	-79	1
-83	0	-88	4	-91	1	-95	0	-103	1
-104	0	-107	0	-115	4	-116	0	-119	0
-127	4	-131	0	-136	4	-139	9	-143	0
-148	1	-151	1	-152	0	-155	0	-163	1
-164	0	-167	0	-179	0	-184	4	-187	4
-191	0	-199	9	-203	0	-211	9	-212	0
-215	0	-223	4	-227	0	-232	16	-235	16
-239	0	-244	9	-247	9	-248	0	-251	0
-259	1	-260	0	-263	0	-271	1	-280	4
-283	0	-287	0	-292	1	-295	16	-296	0
-299	0	-307	0	-308	0	-311	0	-319	4
-323	0	-328	16	-331	9	-335	0	-340	16
-344	0	-347	0	-355	4	-356	0	-359	0
-367	9	-371	0	-376	4	-379	9	-383	0
-388	1	-391	16	-395	0	-403	4	-404	0
-407	0	-415	4	-419	0	-424	0	-427	1
-431	0	-436	0	-439	0	-440	0	-443	0
-451	0	-452	0	-455	0	-463	1	-467	0
-472	36	-479	0	-487	1	-488	0	-491	0
-499	0								

$\Gamma_0(27)$: Special values for

Even quadratic characters

m_χ	$\Lambda(\chi)$	m_χ	$\Lambda(\chi)$	m_χ	$\Lambda(\chi)$	m_χ	$\Lambda(\chi)$	m_χ	$\Lambda(\chi)$
5	0	8	0	13	3	17	0	28	3
29	0	37	3	40	12	41	0	44	0
53	0	56	0	61	3	65	0	73	3
76	3	77	0	85	0	88	12	89	0
92	0	97	3	101	0	104	0	109	0
113	0	124	12	133	3	136	12	137	0
140	0	145	12	149	0	152	0	157	12
161	0	172	0	173	0	181	27	184	12
185	0	188	0	193	3	197	0	205	12
209	0	217	12	220	12	221	0	229	0
232	0	233	0	236	0	241	3	248	0
253	0	257	0	265	12	268	27	269	0
277	12	280	12	281	0	284	0	293	0
296	0	301	12	305	0	313	3	316	27
317	0	328	0	329	0	332	0	337	27
341	0	344	0	349	3	353	0	364	3
365	0	373	3	376	12	377	0	380	0
385	12	389	0	397	0	401	0	409	3
412	27	413	0	421	3	424	0	428	0
433	48	437	0	440	0	445	48	449	0
457	0	460	12	461	0	469	27	472	12
473	0	476	0	481	3	485	0	488	0
493	12	497	0						

$\Gamma_0(32)$: Special values for

Odd quadratic characters

m_χ	$\Lambda(\chi)$	m_χ	$\Lambda(\chi)$	m_χ	$\Lambda(\chi)$	m_χ	$\Lambda(\chi)$	m_χ	$\Lambda(\chi)$
-3	1	-7	0	-11	1	-15	0	-19	1
-23	0	-31	0	-35	4	-39	0	-43	9
-47	0	-51	4	-55	0	-59	1	-67	1
-71	0	-79	0	-83	1	-87	0	-91	4
-95	0	-103	0	-107	9	-111	0	-115	4
-119	0	-123	4	-127	0	-131	9	-139	1
-143	0	-151	0	-155	16	-159	0	-163	9
-167	0	-179	1	-183	0	-187	4	-191	0
-195	16	-199	0	-203	16	-211	1	-215	0
-219	0	-223	0	-227	1	-231	0	-235	4
-239	0	-247	0	-251	9	-255	0	-259	16
-263	0	-267	4	-271	0	-283	9	-287	0
-291	0	-295	0	-299	0	-303	0	-307	25
-311	0	-319	0	-323	0	-327	0	-331	9
-335	0	-339	4	-347	1	-355	16	-359	0
-367	0	-371	0	-379	1	-383	0	-391	0
-395	0	-399	0	-403	4	-407	0	-411	4
-415	0	-419	1	-427	4	-431	0	-435	16
-439	0	-443	25	-447	0	-451	4	-455	0
-463	0	-467	49	-471	0	-479	0	-483	16
-487	0	-491	9	-499	1				

$\Gamma_0(32)$: Special values for

Even quadratic characters

m_χ	$\Lambda(\chi)$	m_χ	$\Lambda(\chi)$	m_χ	$\Lambda(\chi)$	m_χ	$\Lambda(\chi)$	m_χ	$\Lambda(\chi)$
5	0	13	0	17	4	21	0	29	0
53	4	37	0	41	0	53	0	57	4
61	0	65	0	69	0	73	4	77	0
85	0	89	4	93	0	97	4	101	0
105	16	109	0	113	16	129	4	133	0
137	0	141	0	145	0	149	0	157	0
161	0	165	0	173	0	177	4	181	0
185	16	193	4	197	0	201	4	205	0
209	4	213	0	217	16	221	0	229	0
233	4	237	0	241	4	249	4	253	0
257	0	265	0	269	0	273	16	277	0
281	4	285	0	293	0	301	0	305	16
309	0	313	0	317	0	321	4	329	16
337	16	341	0	345	16	349	0	353	0
357	0	365	0	373	0	377	16	381	0
385	16	389	0	393	4	397	0	401	4
409	16	413	0	417	36	421	0	429	0
433	36	437	0	445	0	449	4	453	0
457	0	461	0	465	0	469	0	473	4
481	16	485	0	489	4	493	0	497	16

$\Gamma_0(36)$: Special values for

Even quadratic characters

m_χ	$\Lambda(\chi)$	m_χ	$\Lambda(\chi)$	m_χ	$\Lambda(\chi)$	m_χ	$\Lambda(\chi)$	m_χ	$\Lambda(\chi)$
5	1	13	2	17	3	29	1	37	0
41	3	53	1	61	8	65	0	73	6
77	4	85	2	89	3	97	6	101	9
109	2	113	3	133	8	137	3	145	6
149	1	157	8	161	12	173	1	181	2
185	12	193	0	197	9	205	2	209	12
217	0	221	4	229	18	233	3	241	6
253	8	257	3	265	6	269	9	277	2
281	3	293	1	301	8	305	0	313	24
317	1	329	0	337	6	341	16	349	0
353	27	365	16	373	0	377	12	385	24
389	9	397	8	401	3	409	6	413	16
421	2	433	0	437	4	445	2	449	3
457	6	461	25	469	0	473	.12	481	24
485	0	493	2	497	12				

$\Gamma_0(49)$: Special values for

Even quadratic characters

m_χ	$\Lambda(\chi)$	m_χ	$\Lambda(\chi)$	m_χ	$\Lambda(\chi)$	m_χ	$\Lambda(\chi)$	m_χ	$\Lambda(\chi)$
5	1	8	2	12	4	13	1	17	1
24	4	29	2	33	4	37	2	40	0
41	1	44	0	53	0	57	0	60	8
61	1	65	2	69	4	73	9	76	4
85	2	88	8	89	9	92	8	93	8
97	1	101	1	104	16	109	2	113	8
120	8	124	16	129	16	136	4	137	2
141	8	145	4	149	0	152	4	156	8
157	9	165	8	172	0	173	1	177	32
181	1	184	8	185	16	188	0	193	0
197	0	201	0	204	8	205	2	209	16
213	4	220	16	221	2	229	1	232	8
233	18	236	4	237	0	241	1	248	0
249	8	253	8	257	25	264	0	265	36
268	8	269	9	277	8	281	2	284	32
285	16	293	1	296	8	305	18	309	8
312	8	313	1	316	0	317	8	321	4
328	36	332	4	337	2	341	4	344	8
345	8	348	16	349	9	353	25	365	2
373	0	376	16	377	4	380	8	381	0
389	18	393	8	397	1	401	18	408	8
409	9	412	16	417	8	421	8	424	8
428	8	429	8	433	9	437	0	440	16
444	16	445	18	449	32	453	0	456	32
457	8	460	16	461	1	465	0	472	4
473	8	481	16	485	18	488	0	489	16
492	8	493	0						

§6.3. $X_1(13)$, Odd quadratic characters

The following is a table of special values

$$\Lambda(\chi) = \frac{\tau(\bar{\chi})\, L(f, \chi, 1)}{\Omega_f^-}$$

for primitive odd quadratic characters χ, where f is the normalized cusp form for $\Gamma_1(13)$ with nebentypus character $\epsilon : (\mathbf{Z}/13\,\mathbf{Z})^* \to \mathbb{C}^*$ satisfying

$$\epsilon(2) = \rho = e^{\pi i/3} \quad .$$

These special values satisfy the congruence

$$\Lambda(\chi) \equiv \chi(13) \cdot \epsilon(m_\chi) \cdot \mathbf{B}_1(\chi) \cdot \mathbf{B}_1(\bar{\epsilon}\chi) \pmod{3 + 2\rho}$$

(compare §4.3).

$\chi_1(13)$: Special Values for

Odd quadratic characters

$\Lambda(\chi) = a + b\cdot\omega \; ; \; \omega = (1+\sqrt{-3})/2$

m_χ	$a + b\cdot\omega$	m_χ	$a + b\cdot\omega$
3	2, -2	5	0, -2
7	0, -2	11	6, -6
13	10, -10	17	8, -8
19	0, -14	23	2, -2
29	0, 0	31	-4, 0
37	-4, 0	41	12, 0
43	0, -22	47	4, 0
53	0, 10	59	0, -2
61	-4, 4	67	38, -38
71	0, -2	73	0, -38
79	-8, 0	83	28, 0
89	6, 0	97	54, -54
101	20, 0	103	-4, 0
107	14, -14	109	0, -56
113	8, -8	127	10, -10
131	-4, 0	137	52, -52
139	0, -50	149	56, -56
151	-8, 0	157	0, -98
163	0, -50	167	2, -2
173	2, -2	179	14, -14
181	0, -62	191	0, -10
193	-8, 0	197	46, 0
199	0, -6	211	22, -22
223	18, -18	227	0, -10
229	0, -168	233	0, -8
239	8, 0	241	4, -4
251	0, -18	257	16, 0
263	6, -6	269	34, -34
271	14, -14	277	20, -20
281	0, -2	283	54, -54
293	52, -52	307	48, 0
311	0, 0	313	0, -112
317	0, 22	331	0, -74
337	0, -94	347	0, 10
349	-82, 0	353	80, 0
359	0, 0	367	14, -14
373	28, -28	379	58, -58
383	0, 6	389	0, -46
397	158, -158	401	36, 0
409	40, -40	419	46, -46
421	0, -134	431	6, -6
433	116, -116	439	2, -2

$X_1(13)$: Special Values for

Odd quadratic characters

$\Lambda(X) = a + b\cdot\omega$; $\omega = (1+\sqrt{-3})/2$

m_X	$a + b\cdot\omega$	m_X	$a + b\cdot\omega$
443	52, 0	449	14, -14
457	20, 0	461	32, -32
463	-8, 0	467	16, 0
479	14, -14	487	0, -2
491	70, -70	499	-52, 0
503	0, -14	509	88, 0
521	0, -58	523	22, -22
541	0, -94	547	8, 0
557	76, 0	563	0, 2
569	14, 0	571	-128, 0
577	0, -26	587	18, -18
593	0, 66	599	8, 0
601	-40, 0	607	0, -14
613	144, 0	617	76, -76
619	-116, 0	631	0, -26
641	12, -12	643	0, -122
647	2, -2	653	22, 0
659	0, -54	661	16, 0
673	-40, 0	677	0, 8
683	0, 18	691	90, -90
701	0, -72	709	84, -84
719	0, 2	727	-44, 0
733	0, -206	739	150, -150
743	14, -14	751	-30, 30
757	-62, 0	761	42, -42
769	-122, 0	773	38, -38
787	0, -106	797	58, -58
809	28, 0	811	-96, 0
821	50, 0	823	0, -50
827	44, 0	829	-40, 0
839	0, -18	853	0, -150
857	0, -18	859	-124, 0
863	8, 0	877	88, -88
881	-84, 0	883	84, 0
887	18, -18	907	18, -18
911	-4, 0	919	0, -38
929	44, -44	937	0, -178
941	0, -24	947	10, -10
953	82, -82	967	24, 0
971	0, -26	977	92, 0
983	4, 0	991	2, -2
997	-24, 24	1009	0, -248
1013	0, -18	1019	40, 0
1021	148, -148	1031	0, -6

$\chi_1(13)$: Special Values for

Odd quadratic characters

$\Lambda(\chi) = a + b\cdot\omega$; $\omega = (1+\sqrt{-3})/2$

m_χ	$a + b\cdot\omega$	m_χ	$a + b\cdot\omega$
1033	384, -384	1039	-16, 0
1049	84, -84	1051	42, -42
1061	0, -114	1063	38, -38
1069	-18, 0	1087	-4, 0
1091	-8, 0	1093	0, -74
1097	0, 16	1103	22, -22
1109	32, -32	1117	0, -224
1123	-20, 0	1129	20, 0
1151	0, -26	1153	96, -96
1163	0, -6	1171	-144, 0
1181	98, 0	1187	0, -62
1193	10, 0	1201	0, -166
1213	100, -100	1217	0, 118
1223	4, 0	1229	76, -76
1231	0, -26	1237	-8, 0
1249	0, -282	1259	158, -158
1277	56, 0	1279	-88, 0
1283	0, 54	1289	-16, 0
1291	0, -178	1297	12, 0
1301	0, 24	1303	6, -6
1307	0, 30	1319	0, 2
1321	0, -146	1327	4, 0
1361	16, -16	1367	26, -26
1373	0, 88	1381	64, 0
1399	-28, 0	1409	0, 2
1423	0, -42	1427	22, -22
1429	0, -54	1433	72, 0
1439	0, -14	1447	0, -42
1451	16, 0	1453	-96, 0
1459	34, -34	1471	42, -42
1481	0, -14	1483	40, 0
1487	52, 0	1489	10, -10
1493	152, 0	1499	0, -10
1511	14, -14	1523	42, -42
1531	22, -22	1543	0, -38
1549	-116, 0	1553	20, -20
1559	0, 0	1567	0, -42
1571	78, -78	1579	0, -266
1583	22, -22	1597	118, 0
1601	-30, 0	1607	16, 0
1609	-224, 0	1613	0, -72
1619	0, 22	1621	-126, 126
1627	298, -298	1637	0, 0
1657	356, -356	1663	-24, 0

$X_1(13)$: Special Values for

Odd quadratic characters

$\Lambda(\chi) = a + b\cdot\omega$; $\omega = (1+\sqrt{-3})/2$

m_χ	$a + b\cdot\omega$	m_χ	$a + b\cdot\omega$
1667	62, -62	1669	0,-126
1693	32, 0	1697	20, -20
1699	0,-218	1709	74, -74
1721	0, -90	1723	0,-262
1733	-8, 8	1741	0,-110
1747	-32, 0	1753	-52, 0
1759	0, -6	1777	170,-170
1783	82, -82	1787	0, 82
1789	0,-158	1801	504,-504
1811	0, -86	1823	22, -22
1831	-38, 38	1847	28, 0
1861	-76, 0	1867	-164, 0
1871	-12, 0	1873	0,-206
1877	0, 90	1879	0, -46
1889	84, -84	1901	-26, 0
1907	0, -46	1913	188, 0
1931	0,-150	1933	82, -82
1949	0,-210	1951	-68, 0
1973	116, 0	1979	58, -58
1987	542,-542	1993	108,-108
1997	0, 2	1999	-34, 34
2003	8, 0	2011	0, -46
2017	-52, 0	2027	24, 0
2029	0, 22	2039	50, -50
2053	0,-218	2063	0, 2
2069	114, 0	2081	0, -94
2083	138,-138	2087	0, -2
2089	104,-104	2099	0, -38
2111	4, 0	2113	438,-438
2129	88, 0	2131	-128, 0
2137	0,-304	2141	60, -60
2143	118,-118	2153	0, 8
2161	-320, 0	2179	-228, 0
2203	0, -6	2207	10, -10
2213	16, 0	2221	-412, 0
2237	0, 26	2239	-30, 30
2243	0, 58	2251	-10, 10
2267	240, 0	2269	28, -28
2273	202, 0	2281	-132, 132
2287	-52, 0	2293	0, 32
2297	16, -16	2309	0, -62
2311	-18, 18	2333	34, -34
2339	0, 0	2341	0,-266

$\chi_1(13)$: Special Values for

Odd quadratic characters

$\Lambda(\chi) = a + b\cdot\omega \; ; \; \omega = (1+\sqrt{-3})/2$

m_χ	$a + b\cdot\omega$	m_χ	$a + b\cdot\omega$
2347	0, 18	2351	18, -18
2357	282, -282	2371	128, 0
2377	418, 0	2381	76, 0
2383	0, 2	2389	-336, 0
2393	0, -34	2399	0, -34
2411	0, -138	2417	0, 106
2423	4, 0	2437	0, 0
2441	0, 0	2447	2, -2
2459	38, -38	2467	190, -190
2473	58, 0	2477	38, -38
2503	0, -26	2521	0, -26
2531	0, -74	2539	0, -86
2543	44, 0	2549	0, -14
2551	-46, 46	2557	288, -288
2579	12, 0	2591	0, -38
2593	642, -642	2609	32, -32
2617	40, -40	2621	0, -24
2633	-64, 64	2647	24, 0
2657	0, -32	2659	0, -98
2663	10, -10	2671	0, -114
2677	0, -200	2683	0. 0
2687	0, 10	2689	-204, 0
2693	204, 0	2699	68, 0
2707	282, -282	2711	0, -38
2713	462, -462	2719	2, -2
2729	0, -24	2731	-172, 0
2741	172, 0	2749	294, -294
2753	202, 0	2767	6, -6
2777	0, -58	2789	-16, 16
2791	0, -50	2797	86, 0
2801	64, -64	2803	12, 0
2819	238, -238	2833	0, 34
2837	116, 0	2843	0, -82
2851	0, -250	2857	-104, 0
2861	0, -98	2879	0, -38
2887	-44, 0	2897	214, 0
2903	0, -6	2909	-72, 0
2917	0, -242	2927	6, -6
2939	112, 0	2953	-106, 0
2957	152, -152	2963	108, 0
2969	0, -126	2971	0, -374
2999	0, -10	3001	-324, 0

BIBLIOGRAPHY

[1] A. Atkin and J. Lehner, "Hecke operators on $\Gamma_0(m)$," Math. Ann. 185 (1970), 134-160.

[2] B. J. Birch, "Elliptic curves over $\mathbb{Q}$: a progress report," 1969 Number Theory Institute, AMS Proc. Symp. Pure Math. XX (1971), 396-400.

[3] Z. I. Borevich, and I. R. Shafarevich, Number Theory, Academic Press (1966).

[4] H. Davenport, Multiplicative Number Theory, Markham Publishing Co. (1967).

[5] U. Dieter, "Das Verhalten der Kleinschen Funcktionen $\log \sigma_{g,h}(\omega_1, \omega_2)$ gegenüber Modultransformationen und verallgemeinerte Dedekindsche Summen," J. f. d. reine u. angew. Math. 201 (1959), 37-70.

[6] V. G. Drinfeld, "Two theorems on modular curves," Functional Analysis and its Applications 7 (1973), 155-156.

[7] E. Friedman, "Ideal class groups in basic $\mathbb{Z}_{p_1} \times \ldots \times \mathbb{Z}_{p_n}$ extensions with abelian base fields," to appear.

[8] D. Goldfeld, and C. Viola, "Mean values of L-functions associated to elliptic, Fermat and other curves at the centre of the critical strip," Journal of Number Theory 11 (1979), 305-320.

[9] B. Gross, Arithmetic on Elliptic Curves with Complex Multiplication, Springer Lect. Notes in Math. 776 (1980).

[10] ______, "On the factorization of p-adic L-series," Inv. Math. 57 (1980), 83-95.

[11] H. Hasse, Über die Klassenzahl Abelscher Zahlkörper, Akademie-Verlag (1952).

[12] E. Hecke, "Über die Bestimmung Dirichletscher Reihen durch ihre Funktionalgleichung," Math. Ann. 112 (1936), 664-699. = Werke, 591-626.

[13] ______, "Bestimmung der Klassenzahl einer neuen Reihe von algebraischen Zahlkörpern," Nachr. Ges. d. Wiss. Göttingen (1921), 1-23. = Werke, 290-312.

[14] ______, Darstellung von Klassenzahlen als Perioden von Integralen 3. Gattung...," Abh. Math. Sem. Univ. Hamburg 4 (1925), 211-223. = Werke, 405-417.

[15] ______, "Über Modulfunktionen and die Dirichletschen Reihen mit Eulerscher Produktentwicklung. I, II" Math. Ann. 114 (1937), 1-28. = Werke, 644-707.

[16] ______, "Theorie der Eisensteinschen Reihen höherer Stufe und ihre Anwendung auf Funktionentheorie und Arithmetik," Abh. Math. Sem. Univ. Hamburg 5 (1927), 199-224. = Werke, 461-486.

[17] S. Kamienny and G. Stevens, "Special values of L-functions attached to $X_1(N)$," to appear.

[18] K. Iwasawa, Lectures on p-adic L-functions, Ann. Math. Stud. 74 (1972).

[19] D. Kubert, "The universal ordinary distribution," Bull. Soc. Math. France 107 (1979), 179-202.

[20] D. Kubert and S. Lang, "Distributions on toroidal groups," Math. Zeit. 148 (1976), 33-51.

[21] ______, Modular Units, to appear.

[22] S. Lang, Introduction to Modular Forms, Springer-Verlag, Grundl. d. Math. Wiss. 222 (1976).

[23] W. Li, "New forms and functional equations," Math. Ann. 212 (1975), 285-315.

[24] J. Manin, "Cyclotomic fields and modular curves," Russian Math. Surveys, vol. 26 no. 6 (1971), 7-78.

[25] ______, "Parabolic points and zeta functions of modular curves," Izv. Akad. Nauk SSSR, vol. 6 no. 1 (1972) AMS translation, 19-64.

[26] B. Mazur, "On the arithmetic of special values of L-functions," Inv. Math. 55 (1979), 207-240.

[27] ______, "Courbes elliptiques et symboles modulaires," Seminaire Bourbaki (June, 1972).

[28] ______, "Modular curves and the Eisenstein ideal," Publ. Math. I.H.E.S. 47 (1977), 33-189.

[29] ______, "P-adic analytic number theory of elliptic curves and abelian varieties over $\mathbb{Q}$," Proceedings of the International Congress of Mathematicians at Vancouver (1974), 369-377.

[30] B. Mazur and P. Swinnerton-Dyer, "Arithmetic of Weil curves," Inv. Math. 25 (1974), 1-16.

[31] C. Meyer, "Über einige Anwendungen Dedekindscher Summen," J. f. d. reine u. angew. Math. 198 (1957), 143-203.

[32] ______, "Die Berechnung der Klassenzahl abelscher Körper über quadratischen Zahlkörpern," Berlin Akademie Verlag (1957).

[33] A. Ogg, Modular Forms and Dirichlet Series, Benjamin (1969).

[34] ______, "Rational points on certain elliptic modular curves," Analytic Number Theory, AMS Proc. Sympos. Pure Math. XXIV (1973).

[35] H. Rademacher, "Zur Theorie der Modulfunktionen," J. f. d. reine u. angew. Math. 167 (1931), 312-336.

[36] ______, Topics in Analytic Number Theory, Grundl. d. Math. Wiss. Band 169 (1973).

[37] K. Rubin, "On the Arithmetic of CM Elliptic Curves in $\mathbb{Z}_p$-extensions," Harvard Ph. D. Thesis (1980).

[38] B. Schoeneberg, Elliptic Modular Functions, Grundl. d. Math. Wiss. Band 203 (1974).

[39] ______, "Zur Theorie der verallgemeinerten Dedekindschen Modulfunktionen," Nachr. Akad. Wiss. Göttingen (1969), 119-128.

[40] ______, "Verhalten von Speziellen Integralen 3. Gattung bei Modultransformationen und verallgemeinerte Dedekindsche Summen," Abh. Math. Sem. Univ. Hamburg 30 (1976), 1-10.

[41] J. P. Serre, "Sur le résidu de la fonction zêta p-adique d'un corps de nombres," C. R. Acad. Sci. Paris (1978), 183-188.

[42] G. Shimura, Introduction to the Arithmetic Theory of Automorphic Functions, Princeton University Press (1971).

[43] ______, "On the periods of modular forms," Math. Ann. 229 (1977), 211-221.

[44] J.-L. Waldspurger, "Correspondence de Shimura," J. Math. pures et appl. 59 (1980), 1-133.

[45] L. C. Washington, "Class numbers and $\mathbb{Z}_p$-extensions," Math. Ann. 214 (1975), 117-193.

[46] J. Weisinger, "Some results on classical Eisenstein series and modular forms over function fields," Harvard Ph. D. Thesis (1977).

[47] E. T. Whittaker, G. N. Watson, A course of Modern Analysis, Cambridge Univ. Press (1940).

Progress in Mathematics
Edited by J. Coates and S. Helgason

Progress in Physics
Edited by A. Jaffe and D. Ruelle

- A collection of research-oriented monographs, reports, notes arising from lectures or seminars
- Quickly published concurrent with research
- Easily accessible through international distribution facilities
- Reasonably priced
- Reporting research developments combining original results with an expository treatment of the particular subject area
- A contribution to the international scientific community: for colleagues and for graduate students who are seeking current information and directions in their graduate and post-graduate work.

Manuscripts

Manuscripts should be no less than 100 and preferably no more than 500 pages in length.

They are reproduced by a photographic process and therefore must be typed with extreme care. Symbols not on the typewriter should be inserted by hand in indelible black ink. Corrections to the typescript should be made by pasting in the new text or painting out errors with white correction fluid.

The typescript is reduced slightly (75%) in size during reproduction; best results will not be obtained unless the text on any one page is kept within the overall limit of 6x9½ in (16x24 cm). On request, the publisher will supply special paper with the typing area outlined.

Manuscripts should be sent to the editors or directly to:
Birkhäuser Boston, Inc., P.O. Box 2007, Cambridge, Massachusetts 02139

PROGRESS IN MATHEMATICS
Already published

PM 1 Quadratic Forms in Infinite-Dimensional Vector Spaces
Herbert Gross
ISBN 3-7643-1111-8, 431 pages, paperback

PM 2 Singularités des systèmes différentiels de Gauss-Manin
Frédéric Pham
ISBN 3-7643-3002-3, 339 pages, paperback

PM 3 Vector Bundles on Complex Projective Spaces
C. Okonek, M. Schneider, H. Spindler
ISBN 3-7643-3000-7, 389 pages, paperback

PM4 Complex Approximation, Proceedings, Quebec, Canada, July 3-8, 1978
Edited by Bernard Aupetit
ISBN 3-7643-3004-X, 128 pages, paperback

PM 5 The Radon Transform
Sigurdur Helgason
ISBN 3-7643-3006-6, 202 pages, paperback

PM 6 The Weil Representation, Maslov Index and Theta Series
Gérard Lion, Michèle Vergne
ISBN 3-7643-3007-4, 345 pages, paperback

PM 7 Vector Bundles and Differential Equations
Proceedings, Nice, France, June 12-17, 1979
Edited by André Hirschowitz
ISBN 3-7643-3022-8, 255 pages, paperback

PM 8 Dynamical Systems, C.I.M.E. Lectures, Bressanone, Italy, June 1978
John Guckenheimer, Jürgen Moser, Sheldon E. Newhouse
ISBN 3-7643-3024-4, 300 pages, paperback

PM 9 Linear Algebraic Groups
T. A. Springer
ISBN 3-7643-3029-5, 304 pages, hardcover

PM10 Ergodic Theory and Dynamical Systems I
A. Katok
ISBN 3-7643-3036-8, 352 pages, hardcover

PM11 18th Scandinavian Congress of Mathematicians, Aarhus, Denmark, 1980
Edited by Erik Balslev
ISBN 3-7643-3034-6, 528 pages, hardcover

PM12 Séminaire de Théorie des Nombres, Paris 1979-80
Edited by Marie-José Bertin
ISBN 3-7643-3035-X, 408 pages, hardcover

PM13 Topics in Harmonic Analysis on Homogeneous Spaces
Sigurdur Helgason
ISBN 3-7643-3051-1, 142 pages, hardcover

PM14 Manifolds and Lie Groups, Papers in Honor of Yozô Matsushima
Edited by J. Hano, A. Marimoto, S. Murakami, K. Okamoto, and H. Ozeki
ISBN 3-7643-3053-8, 480 pages, hardcover

PM15 Representations of Real Reductive Lie Groups
David A. Vogan, Jr.
ISBN 3-7643-3037-6, 771 pages, hardcover

PM16 Rational Homotopy Theory and Differential Forms
Phillip A. Griffiths, John W. Morgan
ISBN 3-7643-3041-4, 264 pages, hardcover

PM17 Triangular Products of Group Representations and their Applications
S.M. Vovsi
ISBN 3-7643-3062-7, 150 pages, hardcover

PM18 Géométrie Analytique Rigide et Applications
Jean Fresnel, Marius van der Put
ISBN 3-7643-3069-4, 215 pages, hardcover

PM19 Periods of Hilbert Modular Surfaces
Takayuki Oda
ISBN 3-7643-3084-8, 133 pages, hardcover

PROGRESS IN PHYSICS

Already published

PPh1 Iterated Maps on the Interval as Dynamical Systems
Pierre Collet and Jean-Pierre Eckmann
ISBN 3-7643-3026-0, 256 pages, hardcover

PPh2 Vortices and Monopoles, Structure of Static Gauge Theories
Arthur Jaffe and Clifford Taubes
ISBN 3-7643-3025-2, 275 pages, hardcover

PPh3 Mathematics and Physics
Yu. I. Manin
ISBN 3-7643-3027-9, 111 pages, hardcover

Printed in the United States
By Bookmasters